SUCCESSIONS OF

MENISCOMYINE AND ALLOMYINE

RODENTS (APLODONTIDAE) IN THE OLIGO-MIOCENE

JOHN DAY FORMATION, OREGON

A contribution of the Museum of Paleontology,
University of California, Berkeley, supported in part by the
Annie M. Alexander Endowment and the Burke Memorial
Washington State Museum, University of Washington.

Successions of Meniscomyine and Allomyine Rodents (Aplodontidae) in the Oligo-Miocene John Day Formation, Oregon

John M. Rensberger

UNIVERSITY OF CALIFORNIA PRESS

Berkeley • Los Angeles • London

UNIVERSITY OF CALIFORNIA PUBLICATIONS IN GEOLOGICAL SCIENCES

Editorial Board: D. I. Axelrod, W. B. N. Berry, R. L. Hay, M. A. Murphy,
J. W. Schopf, W. S. Wise, M. O. Woodburne

Volume 124
Issue Date: July 1983

UNIVERSITY OF CALIFORNIA PRESS
BERKELEY AND LOS ANGELES
CALIFORNIA

UNIVERSITY OF CALIFORNIA PRESS, LTD.
LONDON, ENGLAND

ISBN 0-520-09668-1
Library of Congress Catalog Card Number: 83-1403

Library of Congress Cataloging in Publication Data

Rensberger, John M.
    Successions of meniscomyine and allomyine rodents
(Aplodontidae) in the Oligo-Miocene John Day Formation,
Oregon.

    (University of California publications in geological
sciences; v. 124)
    Bibliography: p.
    Includes index.
    1. Meniscomys. 2. Allomys. 3. Paleontology—
Oligocene. 4. Paleontology—Miocene. 5. Paleontology—
Oregon. I. Title. II. Series.
    QE882.R6R47 1983   569′.323   83-1403

    ISBN 0-520-09668-1

TO MY MOTHER AND FATHER:
SARA CORINNE AND NOAH FRANCIS RENSBERGER

Contents

LIST OF FIGURES

Acknowledgments

The following persons provided access to specimens used in this study, or helpfully discussed related problems: Drs. Lawrence Barnes (Los Angeles County Museum of Natural History), Craig Black (Carnegie Museum), Phil Bjork (South Dakota School of Mines and Technology), William Clemens (UCMP), Mary Dawson (Carnegie Museum), Theodore Downs (Los Angeles County Museum of Natural History), Robert Fields (UM), Morton Green (formerly South Dakota School of Mines and Technology), Joseph Gregory (UCMP), Margarite Hugueney (UL), Malcolm McKenna (AMNH), Mr. Ralph Nichols (UM), Drs. Donald Rasmussen (Davis Oil Company), Donald Savage (UCMP), Arnold Shotwell (formerly UOMNH), Professor Mylan Stout (University of Nebraska State Museum), Drs. Richard Tedford (AMNH), John Wahlert (formerly AMNH), David Whistler (Los Angeles County Museum of Natural History), and John White (ISU). Dr. Margarite Hugueney kindly provided casts of *Allomys ernii* from the Coderet.

I am very greatful to the following persons for their review and most helpful suggestions for part or all of this manuscript: Drs. Black and Rasmussen, Professor Stout, and Drs. Albert E. Wood (Emeritus, Amherst College) and Michael O. Woodburne (University of California, Riverside).

Many persons contributed to the field work which resulted in the new material described in this report, especially Scott Bruce, Nina Jablonsky, Lester Kent, Robert Lawrence, Audrey Mesford, Ben Witte and Beverly Witte. Mike Uhtoff donated to the UWBM an extremely important specimen that he collected privately.

Field work for this project was supported first by the UCMP and later by the UWBM and NSF grant GB-35959. The rendered drawings of bones and teeth were made by Mark Orsen.

Abstract

Early studies of aplodontid rodents from the John Day Formation recognized three genera based on a very small number of specimens of unknown chronologic or phylogenetic relationships to one another. A more intensive search for small mammals over the last 18 years has resulted in the accumulation of a much more complete record of these deposits, especially of *Meniscomys, Allomys* and related genera.

This added breadth of materials and stratigraphic-geographic information permit recognition of distinct species of morphologically similar individuals from restricted chronologic intervals. These species fall into two divergent major groups, one of which (Meniscomyinae) is characterized by a heavy transverse crest extending from the mesostylid to the protoconid, heavy lophs or lophids connecting major cusps, the absence of small accessory crests, and a phyletic trend involving increased hypsodonty. At least three successive species of *Meniscomys* are represented (*M. uhtoffi,* n. sp., *M. hippodus* Cope, and *M. editus,* n. sp.), the last being a massive, lophodont and moderately hypsodont form. From a higher stratigraphic position, a distinct and very brachyodont form, *Rudiomys mcgrewi* (n. gen., n. sp.), lacks the vesicular tympanic septae of *Meniscomys* and shows an incipient bladelike development of the metastylid crest and an arched mesostylid crest characteristic of *Sewelleladon predontia* Shotwell. *Rudiomys mcgrewi* is, however, from deposits much lower stratigraphically than the position of *S. predontia,* which was seemingly derived from high in the *Entoptychus-Gregorymys* Concurrent-range Zone, or the *Mylagaulodon* Concurrent-range Zone.

The hitherto described meniscomyines within and east of the Rocky Mountains represent lineages distinct from *Meniscomys* and the other John Day meniscomyines. All of the John Day meniscomyines are distinct from *Niglarodon* Black. The meniscomyines described from the Lemhi region of Idaho by Nichols (1976), though more hypsodont than *Niglarodon koerneri* from the Deep River Formation, appear to be more closely related to the latter than to *Meniscomys.*

The other major group (Allomyinae) is characterized by the absence of a prominent internal mesostylid crest, the presence of a complete metastylid crest reaching the mesostylid, and a number of small, slender, often short crests directed toward the interior basins of the upper and lower molars. *Allomys* is represented in the stratigraphic collection by four successive species (*A. simplicidens,* n. sp., *A. nitens* Marsh, *A. reticulatus,* n. sp., and *A. tessellatus,* n. sp.), ranging from midway in the *Meniscomys* Concurrent-range Zone to the top of the *Entoptychus-Gregorymys* Concurrent-range Zone. Little change in overall hypsodonty was involved in this succession. Instead, the internal crests became more elongate, higher, tended to subdivide internal space more finely, and eventually united in the terminal species to form numerous discrete fossettids.

The type of *Allomys cavatus* Cope is not matched in its small size or poor development of accessory crests by any specimen in the stratigraphic collection. The position of this species with respect to the vertical morphocline from *A. simplicidens* through *A. tessellatus* would place it below *A. simplicidens.* The deposits immediately beneath the occurrence of *A. simplicidens,* and down to the approximate base of the *Meniscomys* Concurrent-range Zone, yielded a well represented taxon, *Alwoodia magna* (n. gen., n. sp.), which is clearly a distinct lineage and not ancestral to the *Allomys* succession. The teeth of *Alwoodia magna* are larger than those of any species of *Allomys,* although the size ranges of *Allomys nitens* and *Alwoodia magna* overlap. *Alwoodia* exhibits a mixture of primitive and advanced features. Accessory crests are generally absent in the upper cheek teeth; primitive crests in all teeth are much heavier than those in *Allomys* and as high as those in the most advanced species of *Allomys.*

Plesispermophilus ernii Stehlin and Schaub is related to the allomyines but differs from all known North American specimens in certain features, especially the structure of the ectoloph in the upper molars. This relationship is here indicated by placing the species in a distinct genus, *Parallomys* (n. gen.).

Although the vertical successions of both meniscomyine and allomyine species show broad recognizable trends, there are significant differences among several of the stratigraphically adjacent and related groups which suggest that the immediately preceding group did not itself give rise to the superadjacent group. This discordance is most conspicuous in the case of *Alwoodia magna* and the overlying *Allomys simplicidens,* but is also suggested by the slightly discordant morphologic relationships of *Meniscomys hippodus* to *M. editus, Allomys simplicidens* to *A. nitens, A. nitens* to *A. reticulatus* and *A. reticulatus* to *A. tessellatus.*

The two major groups of John Day aplodontids seem to have had rather distantly related origins. The meniscomyines somewhat resemble North American prosciurines in the morphology of the lower molars, whereas the primitive dental condition in the allomyines is more suggestive of the morphology in the European aplodontid, *Plesispermophilus.* It seems likely that the North American Allomyinae is of Old World derivation and shares a close relationship with *Plesispermophilus* and a slightly less close relationship to *Sciurodon* Schlosser. Whether *Meniscomys* is derived from a North American prosciurine is not certain because of the morphologic gap between these groups as presently known.

Introduction

The John Day Formation has produced collections of well preserved fossil vertebrates for over a century. Lack of attention to details of geographic and stratigraphic occurrences in the early days of collecting, and some rather severe problems involved in correlating between outcrops, left the chronologic sequence of the abundant species unknown until recently. Recent studies of the geology of the formation (Hay, 1963; Fisher, numerous studies beginning in 1962) focused attention on some of the physical characteristics of the formation, such as the nonsynchronous character of a prominent vertical transition in lithology and color of the sediments and the distributional relationships of the tuffaceous claystones and tuffs in the fossiliferous parts of the formation.

New collections of fossils acquired commencing in 1961 and continuing up to the present have made possible the recognition of successions of taxa (Rensberger, 1971, 1973; Fisher and Rensberger, 1972). These show that the John Day faunas are a complex series with numerous chronologic changes as well as geographic differences.

The degree of completeness of the long entoptychine succession (Rensberger, 1971) in a thick section of uniform tuffaceous siltstones and claystones of air-fall origin, almost everywhere lacking evidence of rivers or large streams, suggests that that part of the formation contains an exceptionally continuous stratigraphic record of successive terrestrial faunas. The results of the present study show that a similar degree of completeness characterizes the history of successive populations of two additional lineages that occur far beneath the range of *Entoptychus*.

This study presents evidence documenting chronologic and geographic differences and relationships of the two most abundant and time-extensive genera of aplodontid rodents, *Meniscomys* and *Allomys,* together with related genera. *Meniscomys* and its relatives are important because they are probably part of a larger group from which the later Tertiary aplodontoids arose, although the details of this history have yet to be deciphered. *Allomys* and its relatives have the longest recorded chronologic range in the formation among the rodents, and the patterns of morphologic change in this group contrast strongly with those of *Meniscomys.* The relationship of *Allomys* to several European forms may provide an answer to the question of origin of this group.

Previous Work

Marsh (1877) described *Allomys nitens* from the "Upper Miocene of Oregon" and proposed the family Allomyidae for the genus. Cope (1879) proposed two additional

taxa, *Meniscomys hippodus* and *Meniscomys multiplicatus*. Both Marsh and Cope in-
itially believed these taxa to be related to the flying squirrels on the basis of the complex
enamel patterns and the fundamentally sciuridlike shape of the cheek teeth. Cope
(1881) recognized four species, *M. hippodus, M. liolophus, M. cavatus,* and *M. nitens,*
but did not mention *M. multiplicatus,* apparently having concluded that it was a
synonym of *M. nitens.* He also noted a resemblance between *M. hippodus* and
Aplodontia, but still referred the fossil group to the Sciuridae.

Furlong (1910) described *Aplodontia alexandrae,* which provided a morphologic intermediate between the John Day forms and the Recent *Aplodontia rufa.* Miller and
Gidley (1918) proposed the genus *Liodontia* for *Aplodontia alexandrae.* They also
recognized both *Allomys* and *Meniscomys,* and proposed a new genus, *Haplomys,*
based on *M. liolophus.* McGrew (1941) saw a close relationship between *M. cavatus*
and *A. nitens,* and regarded *cavatus* as a species of *Allomys.*

Based on a study of new material from the Coderet quarry, near Branssat, France,
Viret and Casoli (1961) showed that the upper dentition of *Plesispermophilus ernii*
Stehlin and Schaub, 1951, previously represented only by lower teeth, is very similar to
that of *Allomys cavatus* and they referred the European species to *Allomys.* Thaler
(1966:213) noted that *Plesispermophilus argoviensis* Stehlin and Schaub, 1951, from
Kuttigen (Aarau), Switzerland, probably also represents *Allomys,* although it is
represented only by a lower molar. Schmidt-Kittler and Vianey-Liaud (1979) interpreted the material of *P. ernii* and a new related species, *P. macrodon,* as more closely
related to *Plesispermophilus angustidens* Filhol than to *Allomys* and grouped the European species in *Plesispermophilus.*

Additional North American rodents from outside the John Day basin have been
referred to *Meniscomys* and *Allomys* in recent years. *Allomys harkseni* Macdonald,
1963, has been described from several localities in the Monroe Creek Formation of
South Dakota (J. R. Macdonald, 1963; 1970; L. J. Macdonald, 1972) and *Allomys stir-
toni* Klingener, 1968, is based on a single lower molar from the late Miocene Norden
Bridge fauna of Nebraska.

An aplodontid resembling *Meniscomys, Niglarodon koerneri,* was described by
Black (1961) from a specimen collected by Dr. Harold Koerner at a locality near Fort
Logan, Montana. Macdonald (1970:32) referred a specimen from the Sharps Formation, South Dakota, to this species, and Nichols (1976:17) assigned a partial skull and
upper dentition from the Lemhi valley in eastern Idaho to this genus, known otherwise
only from the lower dentition. Nichols (1976:15) also described two new species,
Meniscomys yeariani and *M. petersonensis,* in which he observed a progression in hypsodonty of P_4 upward through the stratigraphic section.

Macdonald (1963:178) referred a heavily worn dentition from the Sharps Formation
to the John Day species, *Meniscomys hippodus.* In a later paper (1970), he described
Meniscomys milleri based on better material from the Sharps Formation. Rensberger
(1980) found structures in *M. milleri* that relate it to *Promylagaulus,* although it
represents a stage of evolution comparable to that of *Meniscomys hippodus,* and proposed for it the genus *Crucimys.*

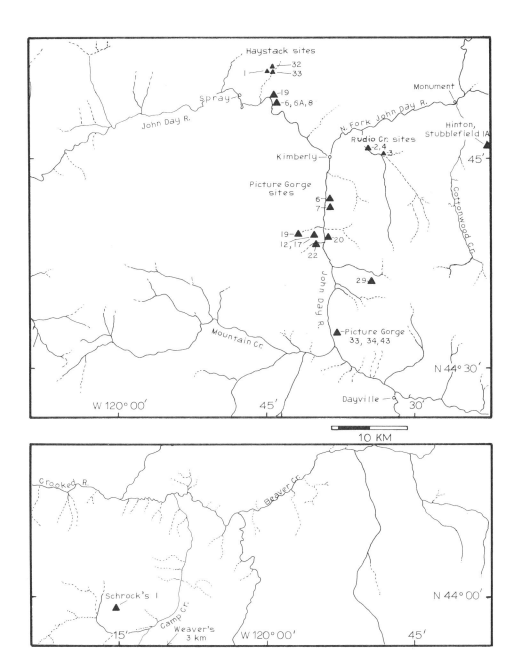

FIG. 1. Maps of localities in eastern Oregon.

Methods of Collection

Almost all of the fossil rodents from the John Day Formation discussed in this study were collected using the same procedures used in earlier studies of the geomyoid rodents (Rensberger, 1971; 1973a). Many were taken during the same years (1961, 1962, 1963), when the field work was directed toward broad and uniform coverage of all outcrops in order to recover as stratigraphically and geographically unbiased a representation of the fauna as possible. However, the lesser abundance of aplodontoids than of geomyoids prompted more selective searches at productive sites and stratigraphic intervals during later years, during which the sample sizes were greatly increased. Each of these specimens was collected either by myself or by a colleague working with me, with the exception of one important specimen, and I accompanied the discoverer and donor of the latter to the position of its recovery. Where specimens were recovered through a significant vertical interval of strata, the interval was either subdivided into smaller units or levels based on distinctive lithologic changes (layering or color differences), or they were allocated to approximate distances above or below the base or top of the outcrop or a distinctive bed.

As with all small mammals from the John Day Formation, most specimens were found as float in the weathering mantle of clay that covers all but the freshest of the exposed rock surfaces. But the large statistical samples of *Entoptychus* (Rensberger, 1971) showed significant morphologic differences in surface collected fossils from successive stratigraphic positions in the same type of rocks and erosional features that produced the aplodontids. Apparently most identifiable specimens are found near their original stratigraphic positions. Those specimens that travel very far do so in gulleys that quickly take them to the bases of the exposures.

Methods of Analysis

Most of the aplodontid specimens recovered may be rather easily allocated to one of two morphologic groups: *Meniscomys* Cope, 1878, or *Allomys* Marsh, 1877.

The specimens of each of these groups have a noticeable range of morphology and frequently the differences in form appear to be associated with differences in stratigraphic or geographic position. Therefore, the specimens allocatable to each of these groups were studied for morphologic characteristics exhibiting associations with stratigraphic level or geographic site, and each specimen was measured or otherwise categorized for each such character. A large list of characters was used for each group, and where there was doubt about the significance of a character, it was included at this stage of the analysis.

When character evaluation was completed for a group, the mean, range, standard deviation and coefficient of variation, together with 95% confidence intervals, were computed for each stratigraphic level (or single level locality). In most instances, owing to the small number of individuals in which a particular character was preserved, differences in computed means from different locality-levels were not statistically significant. However, such differences in means were useful as indications of possible changes.

From these data, pairs of locality-levels with possibly different means were selected and tested under the null hypothesis of independence of frequencies and levels. Fisher's exact test of significance for 2 x 2 contingency tables (Fisher, 1970:96–97) was applied because the resulting likelihood, unlike the X-squared statistic, is exact regardless of sample size, which is usually small in collections of vertebrate fossils partitioned by site and stratigraphic level. The numerical range of values of a variate was divided into two parts, and the number of individuals at each of the two localities or levels falling into each of the two parts of the range formed the four frequencies of a 2 x 2 table.

In the statistical tables, the two morphologic subintervals are presented on the left side of the table, and the two stratigraphic intervals being compared are presented on the right. In effect, the test determines whether a difference in the proportions of individuals falling into the two subintervals are significantly different at the two stratigraphic intervals. The frequencies (numbers of individuals in either of the morphologic intervals) are not listed but may be read from the figured histograms. In a few cases, the class intervals of the histograms do not correspond to the tested intervals and the frequencies read from a histogram may differ by one individual from the figure used in the test.

As an example of interpreting the results shown in a table, see table 2. The likelihood in this example is 0.09 that, if the specimens from locality PG 20, level 2 and those from interval D (PG 7–1, PG 7–2 and PG 17) were derived from the same population, one would obtain from interval D as many as 10 specimens in which the length of M^3 is no greater than 1.76 mm, and none larger, and as many as 1 specimen from PG 20–2 with a length greater than 1.76 mm, and none smaller. The frequencies 10 and 1 are obtained from the histograms of figure 22. In this and other tests, the morphologic range was divided into two intervals to obtain the greatest significance of difference between the localities. Usually this resulted in placing the division at the end of the range at one of the localities.

Measurements of morphologic structures smaller than 2.6 mm were usually made with a Vickers A.E.I. image splitting eyepiece. Larger structures were measured with a microscope reticule.

ACRONYMS OF INSTITUTIONS AND LOCALITY REFERENCES

AMNH	American Museum of Natural History, New York.
ISU	Idaho State University, Pocatello
MV	Prefix of UM locality numbers.
UCMP	University of California Museum of Paleontology, Berkeley.
UL	University of Lyon, France.
UM	University of Montana, Missoula.
UO	Prefix of UOMNH locality numbers.
UOMNH	University of Oregon Museum of Natural History.
UWA, UWB	Prefixes of UWBM locality numbers.
UWBM	University of Washington, Burke Memorial Washington State Museum, paleontological collections.
V–	Prefix of UCMP locality numbers.
YPM	Yale Peabody Museum, New Haven.

Systematic Description

Order Rodentia
Superfamily Aplodontoidea[1] Matthew, 1910
Family Aplodontidae[1] Trouessart, 1897
Subfamily Meniscomyinae (Rensberger, in press)

Subfamily definition. Cheek teeth showing progressive increase in hypsodonty. Mesostyle in upper cheek teeth with moderate labial prominence, closing labial end of central transverse valley; labial surface of ectoloph slightly convex on paracone and metacone; paracone, metacone cusps dominating ectoloph morphology; single cusp in position of metaconule, hypocone absent on posterior cingulum. Metaconid lacking crest joining mesostylid; prominent complete crest running from mesostylid to protoconid; interior space of talonid filled by large, centrally placed mesoconid, large entoconid, large hypoconid, closely set, and crests connecting these cusps; entoconid as posterior as hypoconulid; anteroconid crest absent; interior space of trigonid filled by large, closely set protoconid, metaconid, mesostylid crest. P_4 considerably larger than M_1, with internal metaconid crest joining ectolophid. Incisors with broader faces than in Allomyinae; faces of lower incisors almost flat. Tympanic bulla thin-walled; interior of tympanic cavity subdivided by numerous, closely spaced bony partitions (spaced approximately 0.5 mm or less apart at anteroposterior center of ventral midline where known).

Genera. Meniscomys Cope, 1879, late Olig., N.A.; *Niglarodon* Black, 1961, late Olig., N.A.; *Sewelleladon* Shotwell, 1958, early Mioc., N.A.; *Rudiomys,* n. gen., early Mioc., N.A.

Discussion. The European genus *Ameniscomys* Dehm (1950) is tentatively excluded from the Meniscomyinae because of the following differences: ectoloph less inflected, styles much less prominent; metaconule with primitive size and position, not expanded lingually like a hypocone; protocone, metaconule of P^4 anteroposteriorly elongate, lacking V–shaped lophs; protoconule directly lingual of paracone, not posterolingual; parastyle smaller on P^4, absent on M^{1-3}; molar mesostylid less labially prominent, not connected to metaconid; absence of crest from mesostylid to entoconid; P_4 lacking crest from mesoconid to metastylid crest, with posteriorly directed crest from metaconid to mesoconid. *Ameniscomys* is so unusual that it is perhaps best left *incertae sedis.*

---

[1] With revised definition (Rensberger, 1975b:10–11).

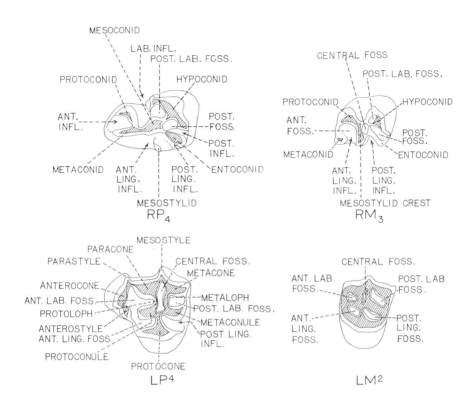

FIG. 2. Nomenclature of dental structures in meniscomyines. Inflections and fossettes(ids) are identified by their position in one of the four quadrants of the occlusal surface: anterolingual, posterolingual, anterolabial or posterlabial; in dead center; or on the boundaries of the quadrants: anterior, posterior, labial or lingual. These structures are homologous in most occurrences among the aplodontoids.

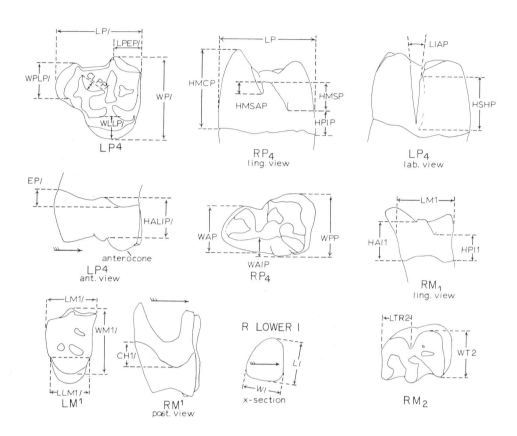

FIG. 3. Measured variates in teeth of meniscomyines. Most variates not shown are similar; see Appendix 1 for complete definitions of both measured and qualitative variates. L^P = length of P^4; LPE^P = length of posterior ectoloph on P^4; W^P = width of P^4; WPL^P = width of parastyle and anterocone on P^4; E^P = depth of lingual enamel on P^4; CLP^P = convexity of paracone on P^4; WLL^P = width of lingual loph of P^4; $HALI^P$ = height of anterolabial inflection on P^4; LM^1 = length of M^1; LLM^1 = lingual length of M^1; WM^1 = width of M^1; CH^1 = posterior chevron height on M^1; LP = length of P_4; HMCP = height of metaconid on P_4; HMSAP = height of mesostylid (anterior) on P_4; HMSP = height of mesostylid (posterior) on P_4; HPIP = height of posterolingual inflection on P_4; WAP = width of anterior moiety of P_4; WPP = width of posterior moiety of P_4; WAIP = width of anterolingual inflection on P_4; LI = length of lower incisor in cross section; WI = width of lower incisor in cross section; LIAP = labial inflection angle on P_4; HSHP = height of stylid on hypoconid of P_4; LM1 = length of M_1; HAI1 = height of anterolingual inflection on M_1; HPI1 = height of posterolingual inflection on M_1; LTR2 = length of trigonid on M_2; WT2 = width of talonid on M_2. Feathered arrows point labiad. The anterior is toward the left in other diagrams.

Meniscomys Cope, 1879

Type species. Meniscomys hippodus Cope, 1879.

Generic definition (revised). Posterolabial fossettid of P_4-M_3 smaller than in *Niglarodon.* Anterolabial inflection open, unlike that of *Sewelleladon;* anterior inflection of P_4 closed toward base, forming fossettid; P_4 trigonid transversely narrower relative to length of tooth than in *Niglarodon.* Metastylid crest on P_4 primitively bladelike but with distinct cusp (mesostylid) at posterior end; in advanced species crest vestigial but mesostylid enlarged, connected to mesoconid by transverse crest. No anterolingual fossettid on P_4, unlike *Sewelleladon.* Mesostylid of molars primitively more bulbous than in *Niglarodon* or *Rudiomys;* mesostylid crest, posterolingual inflection of molars shorter than in *Niglarodon,* not as strongly arched around central fossettid as in *Rudiomys.* Metastylid, metaconid more closely set than in *Rudiomys,* metastylid crest not joined to mesostylid as in *Sewelleladon.* M_{2-3} relatively longer anteroposteriorly, narrower transversely than in *Niglarodon.* Tendency for entoconid of molars to be expanded producing lingual concavity in entoconid crest just anterior to entoconid; entoconid anterolabially elongate, not cylindrical as in *Sewelleladon.* M_2 with large central fossettid or expanded labial end of posterolingual inflection. Posterior inflection of M_3 smaller than in *Niglarodon,* closed at early stage of wear by lingual process from hypoconid region.

P^4 relatively longer anteroposteriorly than in *Niglarodon;* posterolabial, posterolingual fossettes separated from central fossette by anterolabial, anterolingual crests of metaconule; anterolabial crest of metaconule connected to metacone. Posterolabial fossette of M^1 smaller, less persistent with wear than in *Niglarodon.* Upper molars relatively larger anteroposteriorly, narrower transversely than in *Niglarodon.*

Periotic septa reticulate, forming small cells over most of interior of bulla wall and mastoid; cells radially elongate only near external auditory meatus.

Species. M. uhtoffi (n. sp.), late Olig., Ore.; *M. hippodus* Cope, 1879, late Olig., Ore.; *M. editus* (n. sp.), late Olig., Ore.

Discussion. None of the previously described meniscomyine species from the Rocky Mountain or Great Plains regions share the morphologic characteristics definitive of *Meniscomys. Meniscomys milleri* Macdonald, 1970, exhibits characters relating it to *Promylagaulus* (Rensberger, 1980) and is morphologically quite distinct from the taxa in the John Day Formation. *Meniscomys petersonensis* shares more characteristics with *Niglarodon* than with *Meniscomys.* Although in *M. petersonensis* a mesostylid is present on P_4 and is similar to that in the two advanced species of *Meniscomys,* other structures of the premolar are suggestive of derivation from the morphology of *N. koerneri:* large posteriorly open posterior fossettid, anteroposterior breadth of the lingual inflection, breadth and open entrant of the posterolingual inflection. Although a central fossettid is present in M_2 of *M. petersonensis,* the fossettid is smaller than that in *Meniscomys.* Other features of the molars are suggestive of structures in *Niglarodon.*

The P^4 of the type of *Meniscomys yeariani,* treated below as *Niglarodon,* differs from that of *Meniscomys* in the greater size of the posterolingual fossette, the narrowness and shallowness of the central fossette near the mesostyle, and the presence of a connection between the protoconule and the anterior cingulum. These features relate

M. yeariani to *Niglarodon koerneri,* suggesting a closer relationship between these forms than between either of them and *Meniscomys.*

Meniscomys uhtoffi,[2] new species .
(Fig. 4; pls. 1a,c; 2h,i)

Type. Left mandible with I_1 fragment, P_4-M_3; missing anterior end of I_1, protoconid of P_4, entoconid of M_3, angle, coronoid process, condylar process; UCMP 76514.

Type locality. Picture Gorge 12 (V-6685, UWA 9591), 30 m beneath Deep Creek tuff. Also see locality description in Appendix 2.

Stratigraphic distribution. Turtle Cove Member of John Day Formation, from 3 m beneath base of Picture Gorge ignimbrite upward to close beneath Deep Creek tuff. Beneath teilzone of *Meniscomys hippodus.* Lower part of *Meniscomys* Concurrent-range Zone of Fisher and Rensberger (1972). Intervals A, B and C of table 1.

Geographic distribution. Exposures in the drainage of the John Day River in Grant and Wheeler counties, eastern Oregon.

Age. Early Arikareean or late Whitneyan; late Oligocene.

Referred specimens. UWBM 31451, nearly complete skull and mandible, locality Picture Gorge 20 (UWA 4556; V-66114), 4 m beneath Picture Gorge ignimbrite. UWBM 43024, right mandible fragment, P_4, locality Picture Gorge 12 (UWBM A9591; V-6685). UCMP 97066, right mandible, I_1 fragment, P_4-M_3;UCMP 105097, right mandible, I_1 fragment, P_4-M_2; UWBM 39523, right mandible, I_1 fragment, P_4-M_2; all from locality Picture Gorge 22 (V-66116; UWA 5172). UCMP 76996, right mandible, P_4-M_2; Haystack 32 (V-6581). (UCMP 97066, 105097 from close beneath Deep Creek tuff; UWBM 39523 from 24 m beneath Deep Creek tuff.) UWBM 47332, right mandible, P_4-M_2, from top of 1 m thick resistant tuff, Picture Gorge 29, level 2 (UWA 9596).

Diagnosis. Set of prominent grooves present on posterior surface of P^3 (HGP^3); anteroposterior diameter of P^3 (LP^3) greater, height of crown of P^3 (HP^3) less than in other taxa. Transverse width of central fossette on M^3 (WCF^3), anteroposterior length of M^3 (LM^3) greater than in other taxa; M^3 with double mesostyle. Height (0.05 mm) of lingual enamel on P^4 (E^P) low, ratio (0.48) of anteroposterior length (LLM^1) to transverse width (WM^1) of M^1 low. Height of metaconid on P_4 (HMCP) 2.2-2.3 mm, less than in *M. hippodus, M. editus.* Height of posterolingual inflection base on P_4 (HPIP) 0.1-0.7 mm, less than in *M. hippodus, M. editus.*

The following characters in combination are definitive (individually they overlap with some members of *M. hippodus).* Angle between walls of labial inflection of P_4 (LIAP) 12-21 degrees, greater than in *M. hippodus, M. editus.* Height of mesostylid on P_4 (HMSP) 0.0-1.0 mm, height of mesostylid-metastylid notch on P_4 (HMSAP) 0.0-0.5 mm, height of stylid on P_4 hypoconid (HSHP) 0.0-1.1 mm, transverse width of anterolingual inflection on P_4 (WAIP) 0.21-0.46 mm, all less than in *M. hippodus.* Anterior spur (SHLP) seldom present on hypoconulid of P_4; mesostylid process of P_4 (MP) absent. Height of bases of posterolingual inflection (HPI1), anterolingual inflec-

---

[2] Honoring Mr. Mike Uhtoff, who found and donated UWBM 31451, an almost complete skull and jaws of this species.

tion (HAI1) on M_1 less (0.4–0.7 mm, 0.8–1.0 mm, respectively) than in *M. hippodus*. Length (1.6–1.8 mm) of M_1 (LM1) tending to be less than in *M. hippodus*. Width (1.5–1.7 mm) of talonid (WT2), length (0.8–0.9 mm) of trigonid (LTR2) smaller than in *M. hippodus*. Anteroposterior diameter (1.5–1.8 mm) of lower incisor cross section (LI) tending to be smaller than in *M. hippodus*.

Discussion. Most of the characteristics that distinguish *M. uhtoffi* from the stratigraphically overlying form, *M. hippodus,* involve differences in hypsodonty or lophodonty. *M. uhtoffi* consistently ranks lower on both scales. The diagnosis presented does not specifically refer to differences separating *M. uhtoffi* and *M. editus,* but, as the frequency distributions show, the latter represents a morphological extension of most of the trends separating *M. uhtoffi* and *M. hippodus.*

The only known upper dentition of *M. uhtoffi* is that of UWBM 31451, in which the associated lower dentition is very worn and shares few measurable features of hypsodonty or lophodonty with specimens of adjacent beds. The referral of this specimen to *M. uhtoffi* is based upon its morphologic distinction from the other species and its similar stage of advancement and close stratigraphic association with the specimens of *M. uhtoffi.*

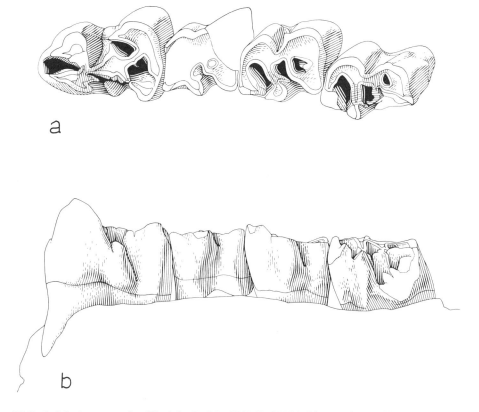

FIG. 4. *Meniscomys uhtoffi,* right P_4-M_3, UCMP 97066, Picture Gorge 22 (V–66116): *a,* occlusal view; *b,* lingual view; anterior left, X10.

Measurements* of Lower Teeth of *M. uhtoffi* (Type)
(UCMP 76514)
(decimal values in millimeters)

LP	WPP	LIAP	HMSP	HMSAP	HMCP	HPIP	HSHP	WAIP
3.05	2.00	20	0.87	0.49	2.21	0.29	0.62	0.25

SHLP	MP	LM1	HAI1	LM2	WT2	LTR2	LI
absent	absent	1.64	0.8	1.86	1.51	0.86	1.71

* See Appendix 1 for definitions of variates.

Meniscomys hippodus Cope, 1879
(Figs. 5–6; pls. 1b–d, 2, 3a–d)

Cope, E. D. 1884, pl. LXIII. McGrew, P. O. 1941, fig. 8a.

Type. Skull and mandible with left P^4-M^2; right P^3-M^3; left P_4-M_1, M_3; right P_4-M_3; AMNH 6962.

Type locality. Drainage of John Day River, Oregon.

Stratigraphic distribution. Turtle Cove Member of John Day Formation. Referred specimens from interval immediately above Deep Creek tuff and beneath teilzone of *M. editus.* Above teilzone of *M. uhtoffi.* Interval D in table 1.

Geographic distribution. Along drainage of John Day River in Grant and Wheeler counties, eastern Oregon.

Age. Early Arikareean; late Oligocene.

Referred specimens. Upper dentitions: UCMP nos. 75508, 75542, 75544, 75545, 75558, 76803, 86269, 86270, 97071, 10545, 105053, 105076–105088, 105090, 107735; UWBM nos. 29205, 29210, 29219, 29224, 29226, 29227, 29229, 29230, 29231–29234, 29284, 29292, 29320, 29323, 29324, 29337, 29341, 29359, 29361, 39514, 39540, 39545, 39546, 39549, 39553, 39554, 39555, 43111, 43125, 43248, 43253, 43257, 43265, 43282, 43283, 43287, 43338, 43340, 43365, 43366, 43371, 43372, 43379, 43385, 43392, 43393, 46965, 47337, 51790. Lower dentitions: UCMP nos. 75546, 75552–75555, 75557, 75559, 76050, 76803, 76996, 86267, 97070, 97072, 105035, 105044, 105058, 105060–105066, 105098–105121; UWBM nos. 29157, 29160, 29170, 29214, 29217, 29223, 29321, 29325, 29340, 29357, 29358, 29360, 29371, 31485, 39507, 39510, 39512, 39544, 39547, 39548, 39551, 39556, 39557, 43254, 43270, 43290, 43337, 43351, 43369, 43370, 43374, 43380, 43388, 43391, 43394, 47330, 48163.

Diagnosis (revised). Height of cusps (for example, metaconid on P_4 [HCMP] 2.5–2.9 mm) greater than in *M. uhtoffi;* less than in *M. editus.* Height of posterolingual inflection base on P_4 (HPIP) 0.5–1.1 mm, greater than in *M. uhtoffi.* Height of stylid on hypoconid of P_4 (HSHP) 0.2–1.6 mm, less than in *M. editus.* Height of base of posterolingual inflection on M_1 (HAI1) 0.8–1.4 mm, greater than in *M. uhtoffi,* less than in *M. editus.* Height of posterolabial inflection base on M_1 (HPI1) 0.3–0.9 mm (except in one unusually large specimen), anteroposterior length of M_1 (LM1) 1.6–2.0 mm, less than in *M. editus.* Height of anterolingual inflection base on M_2 (HAI2) 0.8–1.6 mm, less than in *M. editus.* Height of posterolingual inflection base on M_2 (HPI2) 0.4–0.9 mm, less than in *M. editus.* Anteroposterior length of trigonid on M_2 (LTR2) 40 to 70% of length of M_2 (LM2), greater than in *M. editus.*

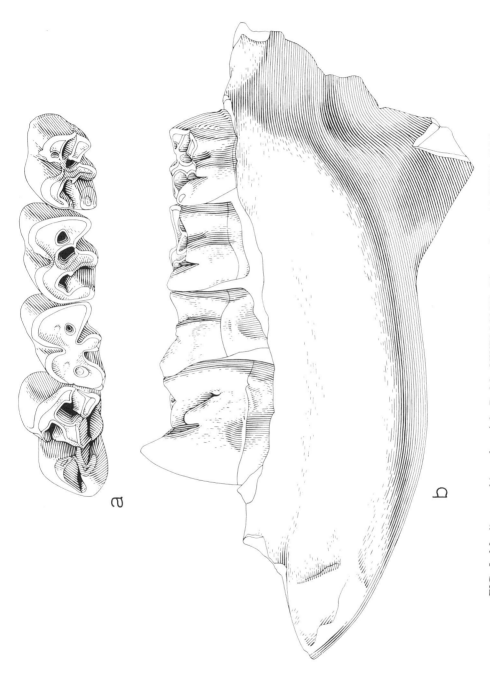

FIG. 5. *Meniscomys hippodus*, right P$_4$-M$_3$, UWBM 39510, Picture Gorge 17 (UWA 5171): *a*, occlusal view; *b*, lingual view; anterior left, X10.

Anteroposterior length of lingual moiety of P^4 (LL^P) 1.60–2.15 mm, anteroposterior length of P^4 (L^P) 3.0–3.8 mm, both less than in *M. editus*. Paracone on P^4 lingually more convex (depth of curvature [CLP^P] 0.15–0.4 mm) than in *M. editus*. Height of anterior enamel above base of anterolabial inflection on P^4 ($HALI^P$) 1.1–1.8 mm, less than in *M. editus*. Height of lingual enamel from apex of posterior enamel-dentine chevron on M^1 (CH^1) 0.3–0.9 mm, height of lingual enamel from apex of chevron on M^2 (CH^2) 0.1–0.7 mm, less than in *M. editus*. Height of lingual enamel on P^4 (E^P) 10 to 25% of length of M^3 (LM^3), 2 to 14% of length of P^4 (L^P), both less than in *M. editus*.

Discussion. A number of characteristics of the holotype fall within ranges of values described for the specimens from interval D and outside the ranges for intervals A, B and C *(M. uhtoffi)* and interval E *(M. editus)* ; for example, HPIP, HAI2, ratio of LTR2 to LM2, LL^P, L^P, and CLP^P (see measurements below).

The specimens in the stratigraphic collections of UCMP and UWBM referred to this species represent considerable morphologic diversity and possibly more than one species population. Some of this diversity results from the inclusion of a few specimens that are intermediate in morphology between *M. hippodus* and *M. editus* but that were recovered from intervals which mainly yielded more primitive morphologies. Several specimens from Picture Gorge 7, level 2, are referred to *M. editus* rather than *M. hippodus* because of pronounced advancement in a number of characters. Discrimination between *M. hippodus* and *M. editus* seems to be optimum when using enamel length measured from the apices of the enamel-dentine chevrons. However, the chevrons occur on anterior and posterior surfaces, which in the specimens with a more complete dentition are hidden by the adjoining teeth and make assessment impossible without damaging the specimen.

The specimen from the Sharps Formation referred to *M. hippodus* by Macdonald (1963:178) could not be found and compared to the John Day taxa. The cheek teeth are apparently heavily worn and the affinity of the specimen may be difficult to determine.

Measurements* of Teeth of *M. hippodus* (Type)
(AMNH 6962)
(decimal values in millimeters)

Lower

LP	WAP	WPP	LIAP	HMSP	HMSAP	HPIP	LM2	WT2
1.6	2.2	2.2	7	1.1	1.2	0.7	1.7	1.5

LTR2	HAI2	WI	LI
0.9	1.6	1.7	1.7

Upper

LP^3	HP^3	LL^P	L^P	CLP^P	WPL^P	W^P	WLLP	LPE^P
1.5	2.07	2.0	3.5	0.2	1.3	3.1	0.6	1.0

LLM^1	LM^2	WM^2	LM^3	W^I	LM^1	WM^1
1.4	1.9	2.3	1.6	1.8	1.7	2.6

*See Appendix 1 for definitions of variates.

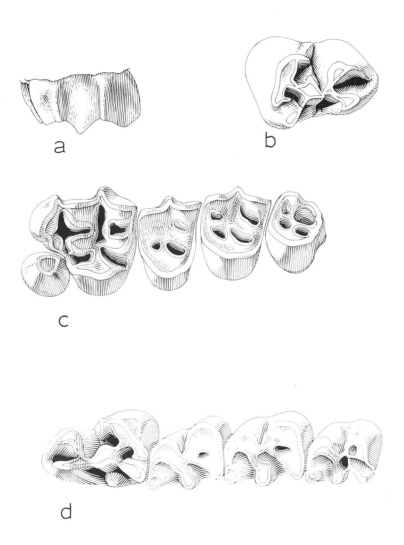

FIG. 6. *Meniscomys hippodus: a,* left P³-⁴, labial view, anterior left, UCMP 76803, Picture Gorge 7, (V–6506), level 2; *b,* left P₄, occlusal view, UCMP 105044, Picture Gorge 7, (V–6506), level 1; *c,* same specimen as *a,* P³-M³, occlusal view; *d,* right P₄-M₃, occlusal view, UCMP 75554, Schrock's 1 (V–6351), level 0; labial top in occlusal views, X10.

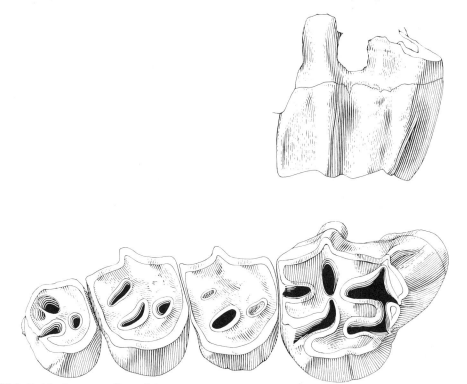

FIG. 7. *Meniscomys editus,* right P⁴-M³, occlusal view, with labial view of P⁴, UWBM 31278, Picture Gorge 7 (UWA 5183), level 3; anterior right, X10.

Meniscomys editus,[3] new species
(Figs. 7–12; pls. 3e, 4)

Type. Fragment of right maxillary, palate, interorbital region, P⁴-M³; UWBM 31278.

Type locality. Picture Gorge 7 (UWA 5183, UCMP V–6506), from middle of level 3 (lower part of interval E). Also see locality description in Appendix 2.

Stratigraphic distribution. Turtle Cove Member of John Day Formation at only known locality from about 17 m above Deep Creek tuff (slightly below level 3 as defined in Rensberger, 1971; fig. 69) to 26 m or more above Deep Creek tuff. Immediately above teilzone of *M. hippodus,* beneath but not reaching teilzone of *Pleurolicus sulcifrons.* Middle part of *Meniscomys* Concurrent-range Zone.

Geographic distribution. Along drainage of John Day River in Grant County, eastern Oregon.

Age. Middle Arikareean; late Oligocene.

Referred specimens. UCMP nos. 97070, left P₄; 105059, right mandible, I₁ fragment, P₄, M₂; 105092, left mandible, I₁ fragment, P₄-M₂; 105075, left maxillary fragment, P₃-P₄. UWBM nos. 31251, left mandible, I₁ fragment, P₄-M₁; 39557, left mandible, P₄-M₃; 54777, left mandible,[4] I₁ fragment, P₄-M₃; 29326, palate, right, left P³-M²;

---

[3] *editus:* L. high, descended.

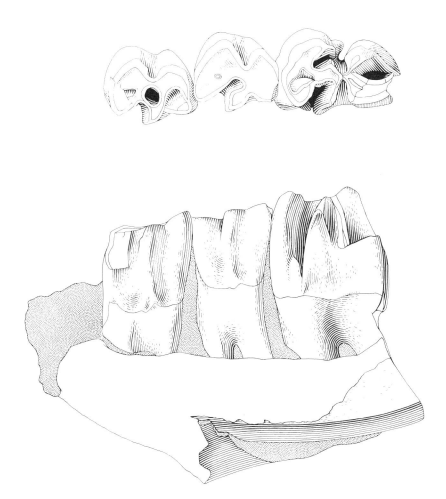

FIG. 8. *Meniscomys editus,* left P$_4$-M$_2$, occlusal (top) and lingual views, UCMP 105092, Picture Gorge 7 (V–6506), level 3; anterior right, X10.

29328 right P^4, 31268, left P^4; 31276, right maxillary, P^4-M^1; 54900, skull,[4] right, left mandible, complete except missing ends of upper and lower incisors, right zygoma, posterior braincase, condylar process of right dentary.

Diagnosis. Height of metaconid on P$_4$ (HMCP) 3.1–3.2 mm, greater than in *M. hippodus.* Height of stylid on hypoconid of P$_4$ (HSHP) >1.9–>3.0 mm, greater than in *M.*

---

[4] Specimens collected after the statistical part of this study had been completed but in which the morphology is like that of the others of this group.

hippodus. Height of base of posterolingual inflection on M_1 (HAI1) 1.54–2.04 + mm, greater than in *M. hippodus.* Height of posterolabial inflection base on M_1 (HPI1) 1.1–1.3 mm, greater than in *M. hippodus.* Anteroposterior length of M_1 (LM1) 2.1–2.2 mm, greater than in *M. hippodus.* Height of anterolingual inflection base on M_2 (HAI2) 1.35–>2.3 mm, greater than in *M. hippodus.* Anteroposterior length of trigonid on M_2 (LTR2) 39 to 42% of length of M_2 (LM2), less than in *M. hippodus.* Height of posterolingual inflection base on M_2 (HPI2) 1.1 mm (2 specimens), greater than in *M. hippodus.*

P^3 with flattened labial surface, less expanded at base than in other species. Anteroposterior length of lingual moiety of P^4 (LL^P) 2.0–2.4 mm, greater than in *M. hippodus.* Anteroposterior length of P^4 (L^P) 3.8–4.3 mm, greater than in *M. hippodus.* Paracone on P^4 lingually flatter (depth of curvature [CLP^P] 0.10–0.15 mm) than in *M. hippodus.* Height of enamel above base of anterolabial inflection ($HALI^P$) 2.0–2.6 mm, greater than in *M. hippodus.* Height of lingual enamel from apex of enamel-dentine chevron on M^1 (CH^1) 2.0 mm, greater than in *M. hippodus.* Height of lingual enamel above apex of posterior chevron on M^2 (CH^2) 1.0–1.4 mm, greater than in *M. hippodus,* height of lingual enamel on P^4 (E^P) 68% of length of M^3 (LM^3), 15 to 27% of length of P^4 (L^P), both greater than in *M. hippodus.*

Discussion. Although represented by a small sample, the specimens from the youngest strata yielding *Meniscomys* are clearly the most advanced in hypsodonty and lophodonty. The teeth in these specimens are also larger than those at lower stratigraphic intervals. Among the referred specimens of *M. editus,* the range in degree of advancement in some characteristics is rather wide, partly because several specimens from level 2 of Picture Gorge 7 (V–6506, UWA 5183) are included in *M. editus.* These specimens are included because in certain characters they are intermediate in morphology between *M. hippodus* and the specimens from level 3, or even as advanced as the latter.

It is possible that these intermediate specimens from level 2 represent a population distinct from either *M. editus* or *M. hippodus,* but the samples studied here are too small to attain significance in statistical tests. However, in UWBM 39557, from high in level 2, the roots of M_1 and M_2 begin to separate at 0.7 mm and 0.4 mm, respectively, below the enamel crown on the lingual side of the tooth, whereas those of UCMP 105092, from level 3, begin to separate at 1.0 mm and 0.9 mm, respectively. In UWBM 31251, from level 3, the roots of M_1 commence separating at 1.1 mm below the crown, and in UCMP 105059 from level 2, the roots of M_2 separate at 0.3 mm. Similar differences in the length of the undivided portion of the root characterize stratigraphically separated stages in the evolution of hypsodonty in the pocket gophers *Entoptychus basilaris* and *E. wheelerensis* in the John Day Formation (Rensberger, 1971; pl. 4a,b). The specimens from the upper part of level 2, although advanced in some aspects of hypsodonty, had not yet begun to acquire unified roots in the lower molars. Larger collections with finer stratigraphic control from the top of level 2 and the base of level 3 at Picture Gorge 7 are needed to resolve these transitional stages in hypsodonty.

The lower dentition of *M. editus* resembles in some degree that of *Niglarodon peter-sonensis.* The metastylid in the molars tapers labially to a rather acute point in both forms, the overall occlusal shape of the molars is similar, and both species are ad-

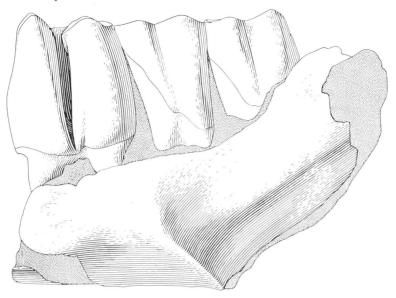

FIG. 9. *Meniscomys editus,* left P$_4$-M$_2$, labial view, UCMP 105092, Picture Gorge 7 (V–6506), level 3; anterior left, X10.

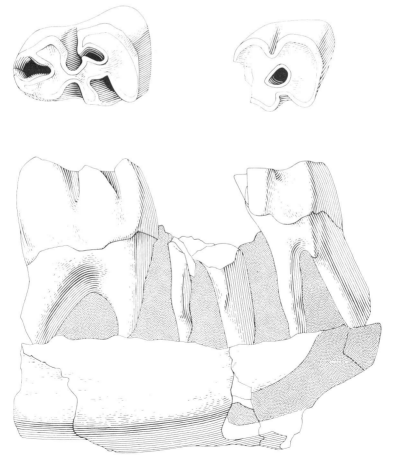

FIG. 10. *Meniscomys editus,* right P$_4$, M$_2$, occlusal (top) and lingual views, UCMP 105059, Picture Gorge 7 (V–6506), level 2; anterior left, X10.

vanced in hypsodonty. However other features of *M. editus* are clearly derived from a morphology like that of *M. hippodus* and are not shared with *N. petersonensis:* reduced M_3; small, closed posterior fossettid of M_3; small posterior fossettid of P_4 and an anteroposteriorly narrow posterolingual inflection of P_4. One feature of *M. editus* that distinguishes it from *N. petersonensis,* the slight flattening of the posterolabial surface of the lower molars, is not shared with *M. hippodus.* This, together with the difference in the shape of the molar mesostylid, suggests derivation from a species of *Meniscomys* not represented in the collection from the John Day and Camp Creek drainages.

Measurements* of Teeth of *M. editus* (Type)
(UWBM 31278)
(decimal values in millimeters)

LL^P	L^P	CLP^P	HALI^P	WPL^P	W^P	WLL^P	LPE^P	LM^1
2.4	4.2	0.1	2.4	1.2	3.7	0.6	1.4	2.5

WM^1	LLM^1	LM^2	WM^2	DMM^3	LM^3	WCF^3	E^P	CH^2
3.2	2.1	2.4	3.0	0	1.7	0.4	1.2	1.0

* See Appendix 1 for definitions of variates.

Niglarodon Black, 1961

Type species. Niglarodon koerneri Black, 1961.

Generic definition (revised). Posterolabial fossettid of P_4-M_3 elongate, larger, more widely open anterolabially than in *Meniscomys.* Metastylid crest, when developed on P_4, lacking distinct cusp (mesostylid) on posterior end; mesostylid, when present, separating anteroposteriorly broader lingual inflections than in *Meniscomys.* Mesostylid present on mesostylid crest of M_1-M_3, but more compressed, less differentiated from crest than in *Meniscomys;* mesostylid crest, posterolingual inflection of molars longer than in *Meniscomys.* Tendency for entoconid crest to maintain almost straight, slightly convex anterolingual outline. Labial end of posterolingual inflection unexpanded or, if central fossettid formed, small. M_3 with large, anteroposteriorly elongate posterior fossettid opening posteriad in upper crown.

Species. Niglarodon koerneri Black, 1961; late Olig., Montana. *N. yeariani* (Nichols), 1976, late Olig., Idaho; *N. petersonensis* (Nichols) 1976, early Mioc., Idaho.

Discussion. Wood (1935) described a mandible from the Deep River Formation bearing a single cheek tooth. Deciduous cheek teeth of aplodontids were not known at that time, and this tooth was interpreted as a permanent M_1 of a new genus of cricetid. Black (1969) recognized the cheek tooth to be a DP_4 and supposed that it belonged to the geomyoid *Dikkomys* Wood (1936), which is also known from the Deep River Formation.

Recently several people, including Black, William Akersten and Donald Rasmussen (personal communication) recognized the tooth as the DP_4 of an aplodontid, possibly *Niglarodon,* which was the only aplodontid that had been described from the Deep River Formation. The UWBM collections then contained a number of distinct and undescribed adult dentitions of aplodontids from the Deep River Formation, which left the affinity of *Horatiomys* in doubt. Wood's description of the type locality of

FIG. 11. *Meniscomys editus,* left P_4-M_1, occlusal (top) and lingual views, UWBM 31251, Picture Gorge 7 (UWA 5183), level 3; anterior right, X10.

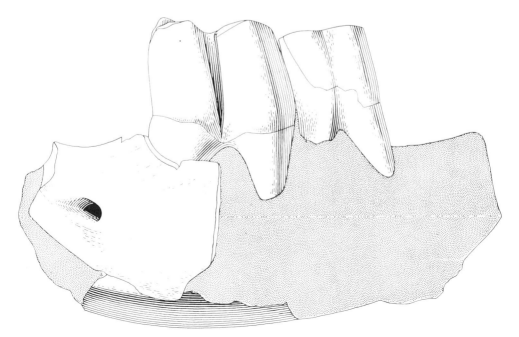

FIG. 12. *Meniscomys editus,* left P_4-M_1, labial view, UWBM 31251, Picture Gorge 7 (UWA 5183), level 3; anterior right, X10.

Horatiomys montanus matches that of a UWBM locality which is sufficiently isolated from other sites to convince me that it represents the *Horatiomys* site. The only aplodontoid taxon in the UWBM collection (and seemingly the most common rodent) from that site, Thompson Gulch North (UWA 8867), is an aplodontid representing the morphology of *Niglarodon koerneri*.

The type of *Horatiomys montanus* may therefore represent an immature individual of the species represented by the type of *Niglarodon koerneri*. However, the DP$_4$ of *Horatiomys montanus* is inadequate for distinguishing species of meniscomyines because the tooth is seldom collected and has not yet been found in the Deep River Formation associated with permanent teeth. The UWBM aplodontoid collections from the Deep River Formation are being described separately.

Although *Niglarodon koerneri* is morphologically similar in many respects to *Meniscomys,* especially primitive members of the latter, the characteristics listed in the revised definition differ consistently from the morphologies of all meniscomyine taxa known from the John Day Formation.

Niglarodon koerneri Black, 1961

(Pl. 5b-d)

Horatiomys montanus Wood, 1935; fig. 2.
Niglarodon koerneri Black, 1961; fig. 1.

    Type. Right mandible, P$_4$-M$_3$; YPM 14024.

    Type locality. Sec. 4, T. 10 N., R. 5 E., Meagher County, Montana.

    Stratigraphic distribution. Deep River Formation.

    Geographic distribution. Exposures along Smith River, Meagher County, Montana.

    Age. Early Arikareean based on stage of hypsodonty compared to that of meniscomyines from John Day Formation; late Oligocene.

    Diagnosis. P$_4$ with anteroposteriorly trending metastylid crest, without distinct mesostylid cusp, without crest from mesoconid to metastylid crest, base of anterior inflection on P$_4$ about 0.5 mm above base of enamel; anteroposterior length of P$_4$ smaller relative to transverse width than in other species (ratio 1.4). Mesostylid crest of M$_{1-2}$ with greatest anteroposterior thickness near lingual end, extending linguad of margin of metaconid. M$_{1-3}$ without closed central fossettid, without vertical continuation of anterolingual inflection beneath point of closure of anterior fossettid. Size of cheek tooth series moderately small; anteroposterior length of series in YPM 14024: 8.7 mm, transverse width of trigonid on M$_3$ 1.8 mm; height of posterolingual inflection base on M$_1$ (HPI1) 0.5 mm; height of anterolingual inflection base on M$_1$ (HAI1) 0.9 mm; height of posterolingual inflection base on M$_2$ (HPI2) 0.35 mm.

Niglarodon petersonensis (Nichols), 1976

    Meniscomys petersonensis Nichols, 1976; pl. 2 (fig. 7), fig. 5.

    Type. Left mandible, P$_4$-M$_3$, UM 5194.

    Type locality. Big Wash North, MV 7303 (= ISU 59003). SE ¼, SE ¼, sec. 8, Lemhi quad., Idaho (Nichols, 1976:11).

Stratigraphic distribution. "Peterson Creek beds," buff colored tuffaceous sediment with occasional layers of water laid gray volcanic ash and a few fluviatile conglomerates of small pebbles (Nichols, 1976:11). *N. petersonensis* from about 53 m above base of exposure, 30 m above occurrence of *N. yeariani,* within stratigraphic range of *Promylagaulus lemhiensis* Nichols, 1976, within but near base of occurrence of *Entoptychus.* Near boundary of *Entoptychus-Gregorymys* and *Meniscomys* concurrent-range zones, on basis of joint occurrence of *Promylagaulus* and *Entoptychus* (see Rensberger, 1979, for other instances of associations or stratigraphic proximity of entoptychines and *Promylagaulus).*

Geographic distribution. Along Lemhi River, Idaho, near Montana border.

Age. Late Arikareean; early Miocene.

Diagnosis. P_4 lacking metastylid crest, with mesostylid connected to mesoconid; apex of metastylid more than 2.8 mm above base of enamel, higher than in *N. yeariani;* base of anterior inflection on P_4 1.9 mm above base of enamel, higher than in *N. yeariani;* posterior inflection between entoconid and hypoconulid open for some vertical extent; hypostylid present on anterior wall of hypoconid region; posterior surface of hypoconid flattened. P_4 longer anteroposteriorly (length 4.1 mm) in relation to transverse width (width of talonid 2.4 mm) than in *N. koerneri* (ratio in *N. petersonensis,* 1.7). M_{1-2} with mesostylid crest reaching maximum anteroposterior thickness near transverse midpoint. M_{1-3} with small, closed central fossettid; with vertical continuation of anterolingual inflection downward, beneath closure of anterior inflection, to base of enamel crown. Size of cheek tooth series, measured at maximum length or width near base of crown, larger than in other species; length of P_4-M_3 11.6 mm; length of P_4 4.1 mm; width of talonid of P_4 2.4 mm; transverse width of trigonid of M_3 2.6 mm. Height of base of anterolingual inflection on M_1 (HAI1) greater than 1.8 mm.

Discussion. UM 5194 is moderately advanced in hypsodonty over YPM 14024, *N. koerneri,* yet it is morphologically closer to the latter than to *Meniscomys* in several characteristics. The molar mesostylid is compressed and poorly differentiated from the crest bearing it. The posterolabial inflection of P_4 opens widely near the upper part of the crown; the central fossettid, although closed on M_2, is small, suggestive of the narrow inflection in *N. koerneri.* Until the upper dentitions of *N. koerneri* and this Lemhi species are described, a provisional assignment of the latter to *Niglarodon* calls attention to a likely affinity between these taxa. The main resemblance to *Meniscomys* is the presence of a distinct mesostylid on P_4, which is lacking in the genotypic species of *Niglarodon.* This structure also occurs in the more primitive meniscomyine from the Lemhi region but, as is shown in the discussion of that species, is not a primitive character for either *Niglarodon* or *Meniscomys* and was independently acquired where it occurs in the two groups.

Niglarodon yeariani (Nichols), 1976

Meniscomys yeariani Nichols, 1976; pl. 2 (figs. 6,8), fig. 5.

Type. Palate with right P^3-M^2, left P^3-M^1, UM 5099.

Type locality. Big Wash, MV 7303. NW ¼ NW ¼ sec. 17, T. 17 N., R. 25 E., Lemhi quad., Idaho.

Stratigraphic distribution. Peterson Creek beds, about 15 m above base of exposure, 55 m below *N. petersonensis,* probably in lower part of *Meniscomys* Concurrent-range Zone (Nichols, 1976:11–13).

Geographic distribution. Along Lemhi River, Idaho, near Montana border.

Age. Early Arikareean; late Oligocene.

Referred specimens. P_4, UM 5012; P_4, UM 5284; both from Big Wash. P_4, UM 4040, from MV 7303, possible landslide.

Diagnosis (revised). P_4 lacking metastylid crest, with metastylid joined to mesoconid; apex of mesostylid about 1.7 mm above base of enamel, lower than in *N. petersonensis;* base of anterior inflection of P_4 0.7–1.0 mm above base of enamel, lower than in *N. petersonensis,* higher than in *N. koerneri;* apex of mesostylid about 1.7 mm above base of enamel, lower than in *N. petersonensis.* Anteroposterior length of P_4 3.2–3.3 mm, greater than in *N. koerneri.* Transverse width of talonid of P_4 1.9–2.0 mm, as in *N. koerneri;* ratio of tooth length to width of talonid 1.7, as in *N. petersonensis* . Other lower teeth of series unknown. P^3 with almost flat labial surface, slightly compressed lingual surface, bladelike posterior carina.

Posterolingual surface of protocone on P^4 sloping more strongly transversely than in *Meniscomys* or *Niglarodon* sp. (UM 5101; Nichols, 1976); opening of posterolingual inflection more widely flaring than in *Meniscomys* or UM 5101; constricton of central valley 0.6 mm labial of bend in posterolingual inflection; labial part of central valley unexpanded (parallel-walled); base of posterolingual inflection about 0.3 mm above base of anterolingual inflection; labial surface of paracone, metacone slightly convex; transverse width of mesostyle 0.4 mm; transverse width of parastyle and anterocone (labial margin of parastyle to notch between anterocone and anterostyle) 1.3 mm, narrower than in UM 5101; anterior moiety of anterolabial fossette transversely narrower than posterolingual inflection.

Height of posterior enamel-dentine chevron on M_1 (CH^1) 0.5 mm; height of posterior chevron on M_2 (CH^2) about 0.8 mm.

Discussion. Although Nichols (1976) chose for the type of *N. yeariani* an upper dentition, whereas both the genotypic species, *N. koerneri,* and the referred species, *N. petersonensis,* are represented only by lower dentitions, he found several lower teeth in stratigraphic association with *N. yeariani.* He stated (p. 13) that *Meniscomys yeariani* is contained in the lower part of the section and (p. 15) that two of the referred lower premolars occurred at the type locality.

The lower premolars, UM 4040 and 5102, although more brachyodont, bear a closer resemblance to the P_4 of *Niglarodon petersonensis* than to that of *N. koerneri.* In both *N. petersonensis* and the P_4's from the lower stratigraphic level there is no metastylid crest, a mesostylid is present, and the ratio of anteroposterior length to transverse width is greater (1.7 compared to 1.4). These differences in morphology represented along the Lemhi River and in the Smith River valley may reflect the isolation of these intermontane basins (about 290 km separation). The differences in known molar series of both basins from those of the John Day basin are consistent with the even greater distance separating the latter from the Lemhi basin (about 470 km).

In assigning this species as well as *M. petersonensis* to *Meniscomys,* Nichols (1976:15) was influenced by the presence of an anterolingual inflection on the P_4 and an elongate metastylid crest, structures that are absent in *N. koerneri*. However, an elongate metastylid crest seems to be primitive for both *Niglarodon* and *Meniscomys* because it is also present in some members of *M. uhtoffi* but is absent in more advanced members of *Meniscomys*.

This resemblance of the lingual morphology of the P₄ in *M. uhtoffi* and *N. montanus* most likely reflects proximity to the common ancestor of the lineages, because these species are the most brachyodont of their respective regions. Divergence from this pattern in *M. hippodus* and *N. yeariani,* both of which are more advanced in hypsodonty, must have been acquired independently. The lower molars of *M. uhtoffi* bear a greater resemblance to those of later species of *Meniscomys* than to those from either of the Rocky Mountain areas. This pattern of similarities and differences is explained if specialization in the lingual structures of P₄ occurred after an early radiation of meniscomyines had transpired, when the molar morphologies distinctive of the different regions had largely been acquired.

? Niglarodon sp.

Niglarodon sp. indet. Nichols, 1976: pl. 2 (figs. 6,8), fig. 6.

Material. Partial skull, I¹, P³-M³, UM 5101.

Locality. Big Wash, MV 7303; (see *N. yeariani).*

Stratigraphic distribution. About 45 m above base of exposure, near occurrence of *Promylagaulus lemhiensis* and short distance beneath occurrence of *N. petersonensis* and *Entoptychus fieldsi* Nichols, 1976 (Nichols, 1976:fig. 2). About 30 m above occurrence of *N. yeariani*. Peterson Creek beds.

Geographic distribution. Along Lemhi River, Idaho, near Montana border.

Age. Middle Arikareean; late Oligocene.

Description. This form is represented by a single specimen, a skull with rostrum rather well preserved, incisors with occlusal ends complete, palatine and maxillary fragments, and complete upper cheek tooth dentition.

P³ is larger (1.9 mm anteroposterior length) than the maximum size in *Meniscomys* (1.6 mm length in *M. uhtoffi),* or the size in the only other described P³ of a meniscomyine (1.5 mm length in *N. yeariani).* The transverse width (1.5 mm) is also greater than in *N. yeariani* (1.3 mm). The greater length of P³ in UM 5101 may reflect larger cheek teeth in general, but only in part. The anteroposterior length of P₄ in *N. yeariani* (3.9 mm) represents 91% of the length (4.3 mm) of P₄ in UM 5101, whereas the length of P³ in *N. yeariani* is only 79% of that in UM 5101. The P³ in UM 5101 also differs from that of other meniscomyines in its occlusal shape. The lingual side is angular, giving the entire occlusal outline a triangular shape. The straight labial side of the triangle reflects a flat labial surface found also in *N. yeariani,* but not in *Meniscomys.*

The P⁴ of UM 5101 is unique among all described aplodontoids in lacking an anterior cingulum. This is so unusual that the possibility of it being an anomalous condition in an otherwise normal population exists. However, the absence of the cingulum functionally correlates with the size and importance of P³ in occlusion, and I suspect the missing cingulum was normal in this form. P³ occupies the notch lingual to the large anterocone of aplodontoids. In advanced forms such as *Promylagaulus,* the region of the anterior cinglum is expanded and reduces the prominence of the notch, with a concomitant reduction of the size and importance of P³, which fails to occlude at all (Rensberger, 1979). In primitive meniscomyines (e.g. *M. uhtoffi),* the notch is prominent, and a large P³ functions as a lingual continuation of the facets on the anterocone of P⁴ for occlusion with the paraconid and metaconid of the lower premolar. The large P³, deep lingual notch, large anterocone, and lack of an anterior cingulum therefore are a functionally related complex that probably characterized this lineage for some time.

The posterolabial fossette of P⁴ in UM 5101 is almost worn away, is small in *N. yeariani,* and larger with regard to the stage of wear in all species of *Meniscomys.*

The labial surface of the paracone on the P⁴ of UM 5101 is almost flat and less convex than in *N. yeariani.* The labial surface of the metacone is concave, whereas that of *N. yeariani* is slightly convex. The paracone is longer anteroposteriorly in both UM 5101 and *N. yeariani* than in *Meniscomys,* giving the P⁴ an asymmetry. The mesostyle is more prominent than in *N. yeariani* or *Meniscomys.*

Metastyles are more prominent near the bases of the crowns of all cheek teeth in UM 5101 than in *N. yeariani* or *Meniscomys.* All molars are larger than in other taxa. M³ is similar in morphology to that in *Meniscomys,* except that the two posterior inflections are more clearly defined (M³ is not known for *N. yeariani).*

Crown height in UM 5101 may be similar, in reference to overall size, to that in the intermediately advanced species of *Meniscomys, M. hippodus,* although enamel-dentine chevrons on the anterior and posterior sides of the teeth could not be measured. The degree of dorsoventral flare of the anterior and posterior sides of the cheek teeth is less than in *N. yeariani,* like that of many specimens of *M. hippodus,* and greater than that in *M. uhtoffi.* The lingual convexity of P⁴ and M² in UM 5101 is less than that in *N. yeariani,* suggesting greater crown height for UM 5101, which is consistent with the higher stratigraphic position of the latter.

Discussion. The labial flatness of P³, the reduction in the size and depth of the posterolabial fossette in P⁴, and the asymmetry of P⁴ resulting from an anteroposteriorly elongate paracone in both UM 5101 and *N. yeariani* suggest a closer relationship between these forms than between either and *Meniscomys.* Until the lower dentition of the group represented by UM 5101 is described, I am regarding it tentatively as *Niglarodon* on the basis of these similarities to *N. yeariani.* It is conceivable that UM 5101 represents the same species as *N. petersonensis* because the occlusal surfaces in those two specimens are of almost the same length, both are unusually large, and their stratigraphic positions at the same locality are close.

Rudiomys,[5] new genus

Type species. Rudiomys mcgrewi, n. sp.

Definition. Periotic with septa enclosing elongate transverse chambers, septa less closely spaced (0.6 mm) than in *Meniscomys;* lateral surface of periotic anterior to auditory tube restricted in area, not depressed as in *Meniscomys;* anteromedial face of

---

[5]Named for Rudio Creek, which runs near the type locality.

periotic less strongly concave than in *Meniscomys;* bulla less inflated than in *Meniscomys.*

Mesostylid of M_{1-2} more bladelike, less swollen than in *Meniscomys;* mesostylid crest of M_2 not straight as in *Niglarodon,* more strongly arched around central fossettid than in *Meniscomys.* Central fossettid of M_2 completely closed. Mesostylid, metaconid more widely separated than in *Meniscomys* or *Niglarodon;* metaconid less inflated than in *Meniscomys* or *Niglarodon,* with apex less labially positioned; metastylid crest more prominent than in *Meniscomys* or *Niglarodon,* with crest extending posterolinguad, not posteriad as in *Crucimys* Rensberger (1980). Anterior fossettid not oval as in *Crucimys,* constricted in center by mesostylid crest. Hypoconid lacking flattening of posterolabial surface characteristic of promylagaulines. Trigonid of M_1 transversely wider relative to talonid than in *Meniscomys* or *Niglarodon.* Posterolingual inflection less compressed, with base extending farther toward center of tooth than in *Meniscomys.*

Molars larger than in other meniscomyines of comparable stage of hypsodonty; absolute crown height as low as in *M. uhtoffi,* perhaps slightly lower than in *Niglarodon koerneri.* Incisor smaller relative to molar size than in other meniscomyines.

Species. Rudiomys mcgrewi n. sp., early Mioc., Ore.

Rudiomys mcgrewi,[6] new species
(Pl. 5a,c,e)

Type. Left mandible, I_1 fragment, M_{1-2}, missing posterior portion of jaw; left periotic; epiphysis of femur; UCMP 105122.

Type locality. Rudio Creek 3 (V–66106), at base of exposure.

Stratigraphic distribution. Type (only known specimen) from Kimberly Member, John Day Formation, probably high in *Meniscomys* Concurrent-range Zone (association with *Allomys nitens;* outcrop has yielded *Promerycochoerus* and not *Entoptychus,* but former from stratigraphically above occurrence of *Rudiomys mcgrewi*).

Geographic distribution. Drainage of North Fork, John Day River, Grant County, Oregon.

Age. Early or middle Arikareean; late Oligocene.

Diagnosis. Only known member of genus.

Discussion. These skeletal parts were found tightly associated in a small concretion. This association together with the fact that the teeth are those of a meniscomyine and the periotic is that of an aplodontid but has finer septal subdivisions than the allomyines indicates that the bones represent a single individual. The absence of vesicular partitioning distinguishes this species from all other meniscomyines in the John Day Formation. Periotics are present in the earliest known specimen of the primitive species, *Meniscomys uhtoffi,* and abundant in the beds yielding *M. hippodus.* The relatively straight, more widely spaced septa radiating from a center in the auditory meatus in *R. mcgrewi* (pl. 5e) is a more primitive condition, judging from the orientation and sparse positioning in *Haplomys,* the earliest known aplodontid in the John

---

[6] Honoring Dr. Paul McGrew and his contribution to the knowledge of North American aplodontoid rodents.

Day Formation. The wide separation of the molar cusps and shape of the metaconid separate *R. mcgrewi* from both *Meniscomys* and *Niglarodon.*

Although aspects of the morphology, including low crown height and less complex bullar septa, suggest primitiveness, the occurrence is stratigraphically higher than the highest known specimens of *Meniscomys,* which leaves little doubt that this represents a lineage distinct from that of *Meniscomys.* Furthermore, the closure of the central fossettid on M_2 is an advanced condition not attained by either *M. uhtoffi* or *Niglarodon koerneri,* but present in advanced species of *Meniscomys.*

That this form is more closely related to the meniscomyine genera than to *Allomys* is shown by the strong mesostylid crest, the absence of an anteroposteriorly aligned crest from the metaconid to the mesostylid, the presence of a central fossettid, and the absence of accessory processes on the major crests. The absence of vesicular partitions in the periotic is shared by *Rudiomys* and *Allomys,* but represents a primitive condition in aplodontids. The bullar septa are more closely spaced than in any of the known periotics of allomyines, including the species that was contemporaneous with this specimen.

Measurements* of Teeth of *R. mcgrewi* (Type)
(UCMP 105122)
(in millimeters)

LM2	WT2	HPI1	HPI2	HAI2	WI	LI
2.2	1.9	0.3	0.4	0.4	1.6	1.9

* See Appendix 1 for definitions of variates.

Sewelleladon Shotwell, 1958

Type species. Sewelleladon predontia Shotwell, 1958.

Definition (revised). Larger than other meniscomyines (18% larger than largest specimen of *Meniscomys editus*) . Lower incisor similar in relative width, but more convex than in *Meniscomys, Niglarodon* or *Rudiomys;* relatively wider, less convex than in *Allomys.*

P_4 with anterolabial inflection closed by process from protoconid to form fossettid; anterior inflection open to base (no anterior fossettid forming with wear); labial face of mesoconid convex, lacking acute process; anterolabial crest of hypoconid connecting lightly to anterior surface of mesoconid; metastylid and mesostylid crests uniting to enclose anterolingual fossettid; crest present on posterior margin of mesostylid partially closing posterolingual inflection to form fossettid.

Lower molars with metastylid crest joining lingual extremity of mesostylid to form flat lingual surface; entoconid cylindrical, not anterolabially elongate in M_{1-2}; posterolingual inflection narrow, transversely short, posterior in position; central fossettid with internally convex labial border (ectolophid), generally triangularly shaped; metaconid anterolingually-posterolabially expanded, with sharp anterolingual stylid. M_{2-3} with crest from mesostylid crest to posterior margin of metaconid forming anterolingual fossettid. Higher crowned than *Rudiomys,* perhaps as hypsodont as *Meniscomys editus.*

Species. Sewelleladon predontia Shotwell, 1958, late early or early middle Mioc., Ore.

Discussion. Although *Sewelleladon* differs from other meniscomyine genera and resembles allomyines in the strong metastylid crest of the molars, some development of the metastylid crest occurs in *Niglarodon*. Other characters indicate a closer relationship to the meniscomyines. The anterior inflection on the P_4 is closed posteriorly by converging crests from the protoconid and metaconid, a characteristic of all meniscomyines but lacking in the allomyines. A complete mesostylid crest is present and shared by all meniscomyines but absent in the allomyines. Three main fossettids (anterior, central, posterior) are present in the molars of this form and are prominent in all meniscomyines but are never distinct in the allomyines. Smaller accessory crests, which are the most distinctive structures in the allomyines, are absent in this specimen as in the meniscomyines. The entoconid is placed at the posteriormost corner of M_{1-2}. None of the allomyines show a tendency toward hypsodonty, whereas in all the meniscomyines and in the related promylagaulines, hypsodonty is an early trend.

A close relationship between *Rudiomys* and *Sewelleladon* may exist. In each, the mesostylid crest bends forward around a large central fossettid, a notch separates the metaconid from the metalophulid I, and in *Rudiomys* the metaconid bears an incipient metastylid crest. Both genera are rare and, as will be shown, probably occur after the disappearance of *Meniscomys,* with the more primitive *Rudiomys* preceding *Sewelleladon.* Evidence for the stratigraphic position of *Rudiomys* is discussed in the treatment of the allomyines.

Sewelleladon predontia Shotwell, 1958
(Pl. 6a,b)

Type. Left mandible, I_1, P_4-M_3; UOMNH F-4734.

Type locality. UO 2275, "in a small north branch of Haystack Creek in section 21, T. 8 S., R. 25 E., three miles northeast of Spray, Oregon" (Shotwell, 1958:452). This is close to locality Haystack 1 (V-6429).

Stratigraphic distribution. Described as "in lower upper John Day Formation" (Shotwell, 1958:452). Up to that time, the "upper John Day Formation" usually referred to the gray colored sediments of the upper part of the formation, following the classification recognized in Merriam's studies (1900, 1901a, 1901b; Merriam and Sinclair, 1907). Such beds overlie greenish sediments of the "middle John Day Formation" at Haystack 1. The color of the type specimen is gray and supports Shotwell's statement of position. The greenish beds of Haystack 1 represent interval F for they contain a fauna of *Entoptychus* less advanced than the species *E. individens* of the Haystack Valley Member, which occurs nearby. However, the gray sediments in this area, as known, represent the Haystack Valley Member, and often those well above the teilzone of *Entoptychus individens* which, as known, occurs in greenish, zeolitized deposits at the base of the member (interval G). Thus the stratigraphic position of *Sewelleladon predontia* is at least as high as the upper part of interval F, and may be as young as or even younger than interval G.

Geographic distribution. Drainage of John Day River, Wheeler County, Oregon. Known only at type locality.

Age. Late Arikareean or early Hemingfordian; early Miocene.

Diagnosis. Only species known.

Subfamily Allomyinae Marsh, 1877

Definition (revised). Cheek teeth brachyodont but with tendencies for development of high crests. Mesostyle in upper cheek teeth with strong labial prominence, closing labial end of central transverse valley; ectoloph with deeply concave labial face on paracone; crests, not cusps, dominating ectoloph; accessory cusp present labial to metaconule, small in P^4, strong in M^{1-2}; hypocone present on posterior cingulum. P_4–M_3 with crest joining metaconid and mesostylid; lower cheek teeth primitively dominated by shallow basin: major cusps small, widely set; hypolophid weak, not continuous across talonid basin; mesostylid primitively a transversely compressed cusp terminating metastylid crest, positioned linguad of line connecting metaconid and entoconid; mesostylid lacking internally directed crest. P_4–M_3 primitively with weak, posteriorly directed crest from anteroconid; crest connecting labial margins of hypoconid and mesoconid closing posterolabial fossettid in most primitive known forms; mesoconid small, strongly labial in position; entoconid anterolingual of hypoconulid; P_4 only slightly larger than M_1, with internal metaconid crest extending posteriad, not toward ectolophid. Incisors with narrower faces than in Meniscomyinae; faces of lower incisors convex.

Interior of tympanic cavity subdivided by sparsely arranged partitions (spaced 2 to 5 mm or more apart at anteroposterior center of ventral midline of bulla).

Genera. Parallomys n. gen., late Olig., Eur.; *Alwoodia* n. gen., late Olig.-early Mioc., N.A.; *Allomys* Marsh, 1877, early-middle Mioc., N.A.; *?Sciurodon* Schlosser, 1884, Olig., Eur.; *?Plesispermophilus* Filhol, 1883, Olig., Eur.

Discussion. Whether *Sciurodon* and *Plesispermophilus* should be included in a broadly defined Allomyinae depends upon the possible derivation of both *Sciurodon* and *Parallomys* from *Plesispermophilus*. The evidence for both possibilities is discussed under the systematics of *Parallomys*. The definition given above is for a more narrowly defined subfamily, although I believe it likely that *Plesispermophilus* and *Sciurodon* will prove to have a close relationship.

Parallomys, new genus

Type species. Parallomys ernii (Stehlin and Schaub), 1951.

Definition. Upper cheek teeth with broad, U-shaped central transverse valley, low protoloph, metaloph. Central, anterior, posterior transverse valleys without accessory crests; walls smooth to faintly crenulated. Labial faces of paracone, metacone sloping strongly linguad; metastylar crest trending strongly labiad.

Lower cheek teeth basined, internal crests low or absent, never reaching center of diagonal valley. Posterointernal crest of mesoconid (ectolophid) not connected to hypoconid.

Species. P. ernii (Stehlin and Schaub), 1951, late Olig., France; *P. argoviensis* (Stehlin and Schaub), 1951, late Olig., Switzerland; *P. macrodon* (Schmidt-Kittler and Vianey-Liaud), 1979, late Olig., Germany.

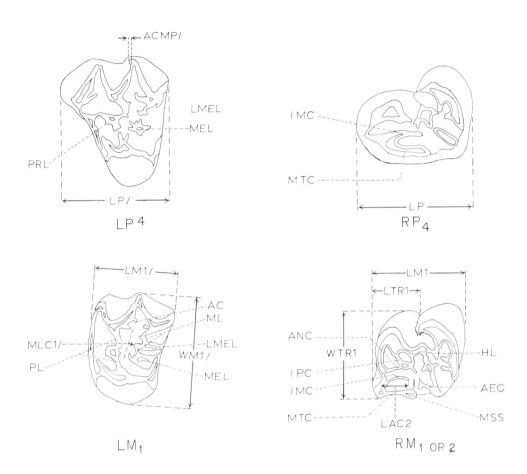

FIG. 13. Measured variates and nomenclature of crests in teeth of allomyines. A number of variates defined for the meniscomyines (see fig. 3) are not repeated here; see Appendix 1 for complete definitions of both measured and qualitative variates. ACM^P = depth of anterior concavity on P^4 mesostyle; L^P = length of P^4; LM^1 = length of M^1; WM^1 = width of M^1; MLC^1 = separation of bend of M^1 metaloph from center of central valley; LP = length of P$_4$; LM1 = length of M$_1$; LTR1 = length of M$_1$ trigonid; WTR1 = width of M$_1$ trigonid; LAC2 = length of anteroconid crest on M$_2$.

Discussion. Stehlin and Schaub (1951) believed their new species from Coderet *(Plesispermophilus ernii)* and from Kuttigen *(Plesispermophilus argoviensis)* were related to *Plesispermophilus angustidens.* Viret and Casoli (1961) recognized the close resemblance of *Plesispermophilus ernii* to *Allomys cavatus* after discovering an upper dentition of *P. ernii.* Later, Thaler (1966: 213), followed by Hugueney (1969), referred the single lower tooth, the type of *Plesispermophilus argoviensis,* to *Allomys.* Schmidt-Kittler and Vianey-Liaud (1979), however, returned to the interpretation of Stehlin and Schaub, that these forms represent *Plesispermophilus,* and believed the similarities to *Allomys* to be due to convergent evolution.

Whether the resemblances between *Parallomys ernii* and *Allomys* resulted from convergence is an issue of importance, because if not convergence, then either the North American or the European form must have originated in the other continent.

Parallomys ernii differs in some features from all North American allomyines. The molars are somewhat transversely wider relative to length than in the American forms. The ectoloph is formed somewhat differently, for it is more deeply inflected and the labial surfaces slope more strongly linguad from base to occlusal edge. A related difference is the longer and more labially trending posterior wing of the loph, the metastylar loph, which forms a more complete W pattern.

Other features of the upper cheek teeth are extremely similar to those of the North American forms, from the small, labially situated second metaconule to the shapes and positions of the other cusps and crests. *Allomys simplicidens* (n. sp.) is closer in structure of the ectoloph to *Parallomys ernii* than is *Allomys cavatus,* which was the form previously compared to the European species. In the type of *Allomys simplicidens,* the paracone and metacone are more deeply inset, almost as much as in *Parallomys ernii* and fully as much as in *Parallomys macrodon.*

Equally convincing of a close relationship between *Parallomys* and *Allomys* is the structure of the lower cheek teeth. The lingual prominence of the mesostylid, the transverse narrowness of the mesostylid, and the development of the metastylid crest in *Parallomys ernii* are essentially indistinguishable from conditions that are found elsewhere only in the primitive North American species, *Allomys cavatus, Allomys simplicidens* and *Alwoodia magna.* The large European species, *Parallomys macrodon,* and the more advanced North American species lack these features. In the North American forms the stratigraphically highest forms depart most strongly from the early condition and the change is incremental.

Although I believe the similarities between the European and North American species are too numerous and complex to be explained by convergence, Stehlin and Schaub's (1951) belief that Mus. Basel Cod. 2180, the type of *Parallomys ernii,* is related to *Plesispermophilus angustidens* (Mus. Basel Q. T. 999) and Schmidt-Kittler and Vianey-Liaud's (1979) similar conclusion are probably correct. In both forms the mesostylid is compressed but is the lingualmost cusp of the lower molars, the lingual margin is convex, the entoconid lies well forward of the posterolingual corner of the molars, the molars are dominantly smoothly basined with only marginal crests, the talonid of M_3 is narrow and posteriorly extended, and the mesostylid is strongly joined to the metastylid crest. Furthermore, *Plesispermophilus* differs in at least several of these characteristics from known primitive nonallomyine aplodontoids of North America, including *Prosciurus, Haplomys, Downsimus* Macdonald (1970), *"Allomys" sharpi* Macdonald (1970) and the meniscomyines. Nevertheless, if *Parallomys* is derived from *Plesispermophilus angustidens,* it has taken a large stride toward the allomyines and deserves separate generic status. *Plesispermophilus* retains the still slightly open posterolabial inflection in M_{1-3} and a low but apparently complete metalophulid II (Stehlin and Schaub, 1951, fig. 468), both of which are found in prosciurines. The structure of the upper and lower dentitions of *Plesispermophilus angustidens,* though suggestive of some conditions in *Parallomys ernii,* probably do not represent the precise ancestry of the allomyines, but may be closer to the ancestry of *Sciurodon cadurcensis* Schlosser, 1884. In the lectotype of the latter, Mus. Basel Q. P. 622 (Stehlin and Schaub, 1951, fig. 489), the entoconid is anteriorly positioned, and the mesostylid is joined to the metaconid by a metastylid crest, as in the allomyines and *Plesispermophilus.* However, like *Plesispermophilus* (especially Mus. Basel Q. U. 273) but unlike the allomyines, the mesoconid is a more distinct cusp more widely separated from the protoconid and a posterolabial crest seems to extend from the protoconid, an unusual feature in the aplodontoids. Both *Sciurodon* and *Plesispermophilus angustidens* have but a single metaconule, unlike the allomyines.

Parallomys ernii (Stehlin and Schaub), 1951

(Pl. 6c-d)

Plesispermophilus ernii Stehlin and Schaub, 1951, fig. 470.
Allomys ernii (Stehlin and Schaub) Viret and Casoli, 1961, figs. 1, 2.

Type. Left mandible, P_4-M_1, Mus. Basel Cod. 2180.

Type locality. Coderet quarry, approximately one km south of Bransat (Allier), France (Viret and Casoli, 1961; Hugueney, 1969).

Stratigraphic distribution. Near the top of massive, somewhat resistant limestone bed immediately overlain by a 1-m-thick argillaceous unit (Hugueney, 1969:5-10).

Age. Late Chattian; Kuttigen Subzone of Coderet Zone (Thaler, 1966:203; Hugueney, 1969:199), late Oligocene if the Oligocene-Miocene boundary is placed at the base of the Aquitanian, as recommended by the Stratigraphic Committee of the Geological Society of London (George, et al., 1969) and the Committee for the Mediterranean Neogene Stratigraphy (Cita, 1968). There are still some unresolved objections to this boundary (summarized though not espoused by Van Couvering, 1972:250).

Referred specimens. Left mandible, P_4-M_2, UL 96303 (fig. 1, Viret and Casoli, 1961); left maxillary, P^4-M^2, UL 96304 (fig. 2, Viret and Casoli, 1961); palate with left P^3-M^2, right P^4-M^3, UL 96305; left mandible, P_4-M_3, UL 96306; right mandible, DP_4-M_2, UL 96307; right maxillary, M^1, UL 96308; DP^4, UL 96309; DP_4, UL 96310; isolated lower teeth: UL 96311-96313; isolated upper teeth: UL 96314-96323.

Diagnosis (revised). Larger than *P. argoviensis,* length of M_1 (LM1) 2.5-2.7 mm. Anteroconid crest curving labiad to join ectolophid at anterior end of mesoconid.

Parallomys argoviensis (Stehlin and Schaub) 1951

Plesispermophilus argoviensis Stehlin and Schaub, 1951, fig. 471. *Allomys argoviensis.* Thaler (1966:213).

    Type. Left M_1, Mus. Basel U.M. 3890.
    Type locality. Kuttigen (Aarau) Switzerland.
    Age. Late Chattian, Kuttigen Subzone of Coderet (Thaler, 1966:203; Hugueney, 1969:199).
    Diagnosis (revised). Smaller than *P. ernii,* length of M_1 1.6 mm. Anteroconid crest curving labiad to join protoconid, like metalophulid II, with branch extending posterolinguad to center of tooth.

Parallomys macrodon (Schmidt-Kittler and Vianey-Liaud), 1979

Plesispermophilus macrodon Schmidt-Kittler and Vianey Liaud, 1979, pl. 1, fig. e.

    Type. Cranial fragment with left P^3-M^3 and right M^{1-3}. Bayer. Staatsslg. Palaontol. hist. Geol. Munchen no. 1952 II-3421.
    Type locality. Gaimersheim, southern Germany.
    Age. Late Oligocene.
    Diagnosis (revised). Larger than other species of *Parallomys.* Paracone, metacone more labially positioned than in *P. ernii.* Lingual metaconule of P^4 less distinct than in *P. ernii.*

Allomys Marsh, 1877

    Type species. Allomys nitens Marsh, 1877.
    Definition. Upper cheek teeth with narrowed central transverse valley; crests of protoloph, metaloph weak or short, walls of central, anterior, posterior transverse valleys with incipient to prominent crests accessory to protoloph, metaloph. Labial faces of paracone, metacone with more vertical attitude than in *Parallomys.* Lower cheek teeth with incipient to prominent crests tending primitively to converge on center of basin but becoming more random in orientation in later forms. Posterointernal crest of mesoconid connecting to hypoconid, enclosing posterolabial fossettid. Molars lingually convex only in primitive species, lingual margin almost straight in later species, with metastylid crest trending posteriad.
    Species. A. cavatus (Cope), 1881, late Olig?, Ore.; *A. simplicidens* n. sp., late Olig., Ore.; *A. nitens* Cope, 1877, late Olig., Ore.; *A. reticulatus* n. sp., early Mioc., Ore.; *A. tessellatus* n. sp., early Mioc., Ore.
    Discussion. These species share the presence of thin, short crests, which in the upper cheek teeth emanate from the crests forming the protoloph and metaloph in primitive aplodontoids. Such structures are essentially absent in the known allomyine specimens from the Great Plains region, in *Alwoodia* from the John Day Formation, and in *Parallomys* from Europe.
    The allomyine specimens described from anthills in the Monroe Creek Formation, South Dakota, by J. R. Macdonald (1963, 1970) and L. J. Macdonald (1972) as *Allomys harkseni* have been recovered in a sample large enough to indicate that more than one taxon is represented. The relationships of these specimens to the allomyines

from the John Day Formation involve quantitative comparisons and will be treated in a separate report.

The type of *Allomys sharpi* Macdonald (1970) is here excluded from *Allomys*. This taxon bears an anterior cingulum on the trigonid of M_{1-3} of the same form as that present in *Downsimus chadwicki* Macdonald (1970). A cingulum in this position is not present in allomyines or other aplodontoids I have seen and therefore suggests a close relationship between these forms. *D. chadwicki,* based on the known material, a mandible, is of prosciurine grade. Both forms lack the anterior placement of the entoconid that distinguishes the primitive allomyines and *Plesispermophilus*. In *A. sharpi* the metaconid is closer to the protoconid than in any of the allomyines. The molars of *A. sharpi,* as expected in primitive allomyines, are basined and lack interior crests, but the marginal cusps are peglike and lack the incipient low crests which descend from the apices toward the basin in both *A. cavatus* and *Parallomys ernii.* In *Plesispermophilus,* which is also of prosciurine grade, the metastylid crest and mesostylid have a morphology more suggestive of that in the primitive allomyines. *A. sharpi* and *Downsimus chadwicki* were probably not members of a lineage leading toward the allomyines. On the basis of the known materials, these species might be regarded as congeneric, because they are similar in size and general morphology and share one unusual feature.

Allomys cavatus (Cope), 1881
(Pl. 7)

Meniscomys cavatus Cope, 1881, pl. LXII, figs. 12–15. McGrew, 1941, fig. 5.

Type. Skull with palate, zygomatic fragment of maxillary, left P^4-M^3, right M^3, right, left periotics; basioccipital, basisphenoid, frontal fragments; left mandible, M_{1-3}; AMNH 6988.

Type locality. John Day region, Oregon. Geographic and stratigraphic position of type unknown, and no referable specimens have been found.

Stratigraphic distribution. Judging from a stage of advancement more primitive than that in *A. simplicidens,* probably middle part of *Meniscomys* Concurrent-range Zone.

Geographic distribution. Type is only known specimen.

Age. Probably late Whitneyan or early Arikareean; late Oligocene.

Diagnosis. Cheek teeth smaller than in other species of genus, width of M^1 (WM^1), 2.6 mm; length of M_1 (LM1) 2.1 mm. One or two low crenulations present on either side of central transverse valley of upper cheek teeth; lower molars with low interior crenulations, including following incipient crests: anteroconid crest, interior protoconid crest (part of metalophulid II), posterolingually and anterolingually directed crests from mesoconid, labially or posterolabially directed crest from entoconid, anterolabially directed crest from entoconid, anterolabially directed crest from hypoconulid, and anterolingually directed crest (ectolophid, in part) from hypoconid to mesoconid. Upper cheek teeth relatively narrower transversely than in *A. simplicidens.* Protoconule, twin metaconules lower, less anteroposteriorly expanded than in *A. simplicidens,* with wear more horizontal and more confined to apices than to sides of cusps. Central valley in upper cheek teeth narrow but not obstructed by crests of protoloph, metaloph; occlusal surfaces of lower molars essentially basins surrounded by cusps with apices positioned at extreme margins of crown. Mesostylid the most lingually situated cusp of lower molars, lingual margins of molars convex, entoconid with anterior position. Metastylid crest only slightly elevated in relief above wall of trigonid basin in earliest stages of wear. Six or seven transverse septa reaching ventral midline of bulla, spaced as much as 4.7 mm apart.

Discussion. This species is morphologically the most primitive of the allomyines known from the John Day Formation. As in *Parallomys ernii,* the lower cheek teeth are basined and only rudimentary interior crests are present. The incipient interior crests on the lower molars and the vertical labial surfaces of the ectoloph in the upper cheek teeth are features shared with the other allomyines in the John Day Formation but not with *Parallomys.*

Allomys simplicidens contains one or two individuals with cheek teeth about as small as those in *A. cavatus* and exhibits the lowest cheek tooth crests in the stratigraphic collection. Nevertheless, *A. simplicidens* is considerably more advanced in crest height and complexity and in flattening of the lingual margin of the lower molars. *A. cavatus* may represent a morphology like that of the ancestor to *A. simplicidens.*

Allomys cavatus is probably not ancestral to *P. ernii.* Incipient crests in the basin of the lower molars and the more vertical labial faces of the paracone and metacone are more advanced than the comparable character states in *P. ernii,* assuming brachyodonty and smooth basins represent the primitive condition. On the other hand, the sloping labial surfaces of the paracone and metacone in *P. ernii* suggest a wider labial shelf than one would expect in an ancestor of *A. cavatus.* It seems probable that *P. ernii* and *A. cavatus* share an origin from some more primitive allomyine.

Measurements* of Dentition of *A. cavatus* (Type)
(AMNH 6988)
(decimal values in millimeters)

Lower

LM1	WM1	WTR1	LAC2	HC	LC	LTR1	NF2
2.1	1.7	1.6	0.71	0	0	1.2	5

Upper

LP^4	LM^1	WM^1	NPC^1	TC^1	NMC^1	$PAPR^1$
2.7	2.2	2.6	2	4	0	1

TC^2	MLC^1	WM^2	ACM^P
7	0.18	2.0	0.0

*See Appendix 1 for definitions of variates.

Allomys simplicidens, new species
(Fig. 14; pl. 8)

Type. Left maxillary with little worn P^3-M^2, UWBM 29156.

Type locality. Picture Gorge 7 (UWA 5183). See detailed locality description in Appendix 2.

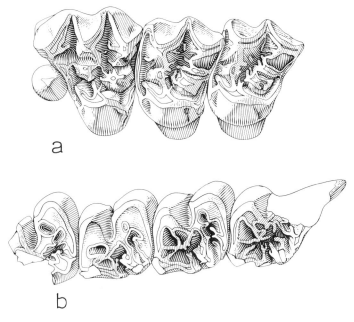

FIG. 14. *a, Allomys simplicidens,* left P³-M², occlusal view, UWBM 29156, Picture Gorge 7 (UWA 5183), level 1; *b, Allomys,* cf. *A. simplicidens,* right P₄-M₃, occlusal view, UCMP 105026, Picture Gorge 7 (V–6506), level 3; anterior left, X10.

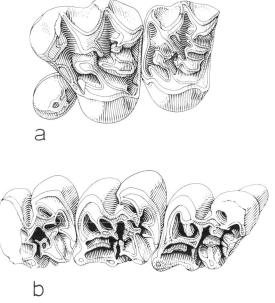

FIG. 15. *Allomys reticulatus: a,* left P³-M¹, occlusal view, UCMP 105033, Haystack 8 (V–6322); *b,* right M₁-₃, occlusal view, UCMP 105034, Haystack 8, level 2; anterior left, X10.

Stratigraphic distribution. Turtle Cove Member of John Day Formation. From immediately above Deep Creek tuff upward through teilzone of *Meniscomys editus.* Middle of *Meniscomys* Concurrent-range Zone.

Geographic distribution. Drainage of John Day River, Grant and Wheeler Counties, Oregon.

Age. Early or middle Arikareean, late Oligocene.

Referred specimens. Picture Gorge 7, level 1 (V–6506, UWA 5183): UCMP 105024, maxillary, M^{1-2}; UCMP 86266, DP^4; UCMP 86268, mandible, P_4-M_1; UWBM 29228, P_4. Picture Gorge 7, level 2: UCMP 97069, maxillary, P^4-M^2; UWBM 54800, mandible, M_{1-2}; UWBM 54801, mandible, P_4-M_3; UCMP 105025, mandible, P_4-M_1; UWBM 29297, mandible fragment, M_1 fragment, M_2. Picture Gorge 7, level 3: UWBM 54786, M^2 fragment; UWBM 31267, M^3; UCMP 105027, M^3; UWBM 54893, P^4 fragment Picture Gorge 29, level 3 (UWA 9596): UWBM 47331, mandible fragment, M_{2-3}. Picture Gorge 33 (V–66104): UCMP 76990, mandible, M_{1-2}.

Diagnosis. Cheek teeth small, but larger than in *A. cavatus:* width of M^1 (WM^1), 2.8–2.9 mm; length of M_1 ($LM1$) 2.1–2.4 mm. Upper cheek teeth relatively wider transversely than in *A. cavatus.* Protoconule and twin metaconules higher, more transversely expanded, forming more persistent protoloph and metaloph than in *A. cavatus,* with anteriorly and posteriorly sloping wear facets meeting symmetrically at apex of cusp. Central valley of upper molars persistently and uniformly V–shaped with wear. Protoloph and metaloph with accessory crests extending anteroposteriorly toward central anterior and posterior transverse valleys; prominence of accessory crests ranging from very low (almost as low as in *A. cavatus)* to moderately low (higher than in *A. cavatus* but lower than in other species from the John Day Formation). Number of accessory crests in DP^4 ($TCDP^4$) 5, fewer than in *A. nitens.*

Interior crests of lower cheek teeth consistently higher than in *A. cavatus,* lower than in *A. nitens;* length of crests shorter than in *A. nitens.* Most fossettids poorly defined after wear, with gently sloping walls. Posterolabial fossettid of molars larger than in *A. cavatus.* Often entoconid positioned more posteriorly, lingual margin of lower molars less convex than in *A. cavatus.* Sides of major cusps facing interior basin steeper than in *A. cavatus.*

Discussion. Allomys simplicidens differs from *Allomys cavatus* mainly in the greater size of the cheek teeth, development of the conules, greater prominence of accessory crests, occurrence of a straighter lingual margin on the lower cheek teeth, and in the position of an entoconid nearer the posterolingual corner of the tooth. In one individual, UWBM 29297, the lingual margin of M_2 is almost as convex as in *A. cavatus,* but the size is larger and the crests are longer and more prominent.

This species is less advanced in crest height than are any of the species occurring at higher stratigraphic positions and is smaller in cheek tooth size than *A. nitens,* which occurs immediately above it.

Measurements* of Dentition of *A. simplicidens* (Type)
(UWBM 29156)
(decimal values in millimeters)

LP^3	WP^3	L^P	LM^1	WM^1	NPC^1	TC^1	NMC^1
1.1	1.0	2.9	2.1	2.9	5	17	6

PAPR^1	TC^2	MLC^1	WM^2	ACM^P
1	14	0.20	2.9	0.0

* See Appendix 1 for definitions of variates.

Allomys nitens Marsh, 1877
(Pls. 9, 10a-c)

Marsh, 1877, fig. p. 253 (crests inaccurately rendered).
Meniscomys multiplicatus Cope, 1879:68
Stehlin and Schaub, 1951, fig. 172 (after Marsh).

Type. Palate with left P^3-M^2, right P^4-M^2, YPM 13604.

Type locality. Somewhere along the John Day River, Oregon.

Stratigraphic distribution. Turtle Cove Member of John Day Formation, from teilzone of *Meniscomys editus* upward to beds near base of *Pleurolicus* teilzone. High in *Meniscomys* Concurrent-range Zone. Above teilzone of *A. simplicidens.*

Geographic distribution. Drainage of John Day River, Grant and Wheeler Counties, Oregon.

Age. Middle Arikareean; late Oligocene.

Referred specimens. Picture Gorge 7, level 3 (V–97074): M^2. Picture Gorge 33 (V–66123, UWA 5833): UWBM 58058, mandible, P$_4$-M$_3$; UCMP 107717, left mandible, DP$_4$-M$_2$, right mandible, P$_4$-M$_2$, left maxillary, DP^4-M^2, right maxillary, DP^4, right premaxillary with I, periotic fragment, humerus fragment, phalanx fragment; UWBM 31518, mandible, M$_1$; UWBM 31498, M$_2$; UCMP 107719, P$_4$; UCMP 107731, mandible, M$_{1-2}$; UWBM 31508, maxillary, M$^{1-2}$; UCMP 107732, P^3; UCMP 107733, upper M fragment Haystack 6 (V–6590, UWA 4799): UCMP 97068, mandible, M$_{1-3}$; UCMP 105032, mandible, M$_{1-3}$; UWBM 28537, lower M fragment; UWBM 28511, maxillary, DP^4-M^2; UWBM 28514, maxillary, P$^{3-4}$; UWBM 28539, M^1; Picture Gorge 19 (UWA 9592): UWBM 43112, maxillary, M$^{1-3}$; UWBM 43107, maxillary, P$^{3-4}$. Picture Gorge 6 (UWA 9581): UWBM 48160, M^1. Rudio Creek 3 (V–66106): UCMP 105041, skull (lacking anterior end of rostrum, jugal, squamosal, braincase), with left P^3-M^3, right M$^{1-3}$, mandible, M$_1$, M$_3$. Picture Gorge 43, level 2 (UWA 5834): UWBM 51967, mandible, P$_4$-M^1; UWBM 31473, mandible, M$_{2-3}$. Picture Gorge 43 (UWA 9966); UWBM 47310, mandible, P$_4$-M$_3$. Merriam's locality 864 (geographical position unknown): UCMP 1444, mandible, P$_4$-M$_3$, skull (lacking anterior end of rostrum, zyomatic arches, palatal bones, basicranial bones) with M^3, fragment P^4-M^2, right bulla, mastoid.

Diagnosis (revised). Cheek teeth larger than in *A. cavatus, A. simplicidens, A. reticulatus,* or *A. tesselatus;* length of M$_1$ (LM1) 2.4–2.6 mm, width of trigonid on M$_1$ (WTR1) 2.0–2.2 mm, length of trigonid on M$_1$ (LTR1) 1.1 -1.3 mm, length of M^1 (LM^1) 2.3–2.7 mm, width of M^1 (WM^1) 3.1–3.5 mm. Width of M^2 (WM^2) 3.2–3.3 mm.

Ratio of length of upper molars to width greater than in *A. simplicidens.* Height of crests greater than in *A. simplicidens.* Number of accessory crests in DP⁴ (TCDP⁴) 15–16, greater than in *A. simplicidens.* Very small accessory crests present, crests less uniform in height, not tending to fill available space as completely as in *A. reticulatus.* Mesostylid more consistently lacking lingual prominence than in *A. simplicidens;* metastylid crest more nearly straight and anteroposteriorly aligned in earliest stages of wear than in *A. simplicidens.* Internal protoconid crest, (labial arm of metalophulid II) on M₂ tending to run straight posterolinguad, or curve slightly anteriad, to join anteroconid crest; fossettid between internal protoconid crest and anterolingual crest of mesoconid, if present, elongate, not oval as in *A. reticulatus.* Anterior accessory crests of paracone and protoconule short, not reaching anterior cingulum in molars; anterior transverse valley unobstructed. Anterior moiety of ectoloph less deeply invaginated basally and adjacent mesostyle less concave anteriorly (ACM^P) than in *A. reticulatus.*

Discussion. The dentition of the type of *A. nitens,* YPM 13604, is smaller than any of the teeth from the localities typifying group E1 (see Sample Relationships of Allomyines) and is intermediate in size between *A. simplicidens* and group E1. However, the length of M¹ (LM¹) in YPM 13604 is closer to the size in group E1 than to that in *A. simplicidens* or *A. reticulatus.* In most of the other characters diagnostic of group E1 and *A. simplicidens,* this specimen is closer to group E1. The ratio of transverse length of M¹ (LM¹) to width of M¹ (WM¹) is 0.77, whereas that in *A. simplicidens* at the typifying locality–levels is 0.68-0.72 and that in group E1 at the typifying localities is 0.76-0.79. Accessory crests in YPM 13604 extend into the central valley, as in group E1, yet the central valley is not as congested as in *A. reticulatus.* The type of *A. nitens* may represent a population distinct from group E1 and somewhat more primitive, but it is more closely related to group E1 than to any other.

UCMP 1444, previously referred to *A. nitens* in the museum documents, fits morphologically in group E1. The periotic region is preserved in this specimen. The bulla contains about 10 internal transverse septa that reach at least to the midline ventrally. Some of these septa originate by branching near the auditory meatus. Only 6 or 7 septa reach the midline in the dentally more primitive species, *A. cavatus.* The dorsal part of the periotic is internally partitioned into small oval cells. The periotic is unknown in the other species.

Measurements* of Dentition of *A. nitens* (Type)
(YPM 13604)
(decimal values in millimeters)

LP³	L^P	LM¹	WM¹	NPC¹	NMC¹	TC¹
1.1	3.3	2.4	3.1	5	6	14

TC²	PAPR¹	MLC¹	WM²	ACM^P
18	1	.22	3.0	0.0

* See Appendix 1 for definitions of variates.

Allomys reticulatus,[7] new species

(Fig. 15; pl. 10d-f)

Type. Palate with right P^3-M^2, left M^1, UCMP 105039.

Type locality. Haystack 6A (V-6505). See detailed locality description in Appendix 2.

Stratigraphic distribution. Turtle Cove and Kimberly members of John Day Formation. Uppermost part of *Meniscomys* Concurrent-range Zone upward through lower part of *Entoptychus-Gregorymys* Concurrent-range Zone. *Pleurolicus, Entoptychus basilaris, E. wheelerensis, E. minor* and *E. cavifrons* teilzones. Above teilzone of *A. nitens.*

Geographic distribution. Drainage of John Day River, Grant and Wheeler counties, Oregon; drainage of Camp Creek, Crook County, Oregon.

Age. Middle to late Arikareean, early Miocene.

Referred specimens. UCMP 105033, maxillary, P^3-M^1, Haystack 8 (V-6322); UCMP 105043, maxillary, M^{1-2}, Stubblefield 1A (V-6658); UCMP 75878, M^1, Schrock's 1 (V-6351), level 8; UCMP 75895, M^1, Schrock's 1, level 8; ?UCMP 76737, M^1 or M^2, Schrock's 1, level 10 (an unusually large specimen); UWBM 39391, M^1 or M^2, Rudio Creek 4 (UWA 5929).

From Schrock's 1, level 8: UCMP 75942, mandible, P_4-M_2; UCMP 75861, mandible, M_{1-3}; UCMP 75894, mandible, M_{1-3}; UCMP 75879, M_2; UCMP 75931, M_1; UCMP 105040, M_1. UWBM 42803, mandible, DP_4 fragment, Picture Gorge 7 (UWA 5183), level 5; UCMP 105037, mandible, M_{2-3}, Haystack 8 (V-6322), level 1; UCMP 105034, mandible, M_{1-3}, Haystack 8, level 2; UCMP 105031, M_2, Haystack 1 (V-6429); UCMP 105042, mandible, M_2, Rudio Creek 4 (V-6600), UCMP 1100, anterior skull fragment, P^3-M^2, Merriam's UCMP locality 851.

Diagnosis. Size of cheek teeth smaller than in *A. nitens,* extensively overlapping size range in *A. simplicidens;* length of M_1, (LM1) 2.0-2.3 mm; width of trigonid on M_1 (WTR1) 1.6-1.9 mm; length of trigonid on M_1 (LTR1) 0.9-1.0 mm; length of M^1 (LM^1) 1.9-2.3 mm; width of M^1 (WM^1) 2.6- 3.0 mm. Accessory crests as abundant as in *A. nitens,* higher than in *A. simplicidens,* tending to fill available space more completely than in *A. nitens;* accessory crests more nearly equal to major crests in size than in *A. nitens.* Accessory crests of upper molars tending to completely cross anterior and central transverse valleys. Anteroposterior lengths of cheek teeth longer than in *A. tessellatus* although width may be similar. Fewer closed fossettids in lower cheek teeth than in *A. tessellatus;* 5-8 fossettids in M_2 (NF2). Length of anteroconid crest on M_2 (LAC2) 0.85-1.05 mm, longer than in *A. tessellatus.* Internal metaconid crest more commonly present and stronger on M_{2-3} than in *A. nitens.* Internal protoconid crest of M_2 usually curving posteriad and forming shorter, more oval fossettid with anterolingual crest of mesoconid. Cingulum at base of labial surface of anterior ectoloph and anterior invagination of mesostyle (ACM^P) more sharply defined than in other species (except unknown in *A. tessellatus).*

Discussion. This species is easily distinguished from all of the foregoing species by the combination of small size and prominence of small crests. It resembles *A. tessellatus* in these characters but lacks the more extreme tendency for union of the

---

[7] *reticulatus:* L. reticulum, net; for netlike, rather even dispersion of fine crests in the lower molars.

crests and for the large number of tiny oval fossettids characteristic of the latter.

The smaller size of *A. reticulatus* and the tendency for anteroposteriorly shorter, relatively wider upper cheek teeth suggests it may not be derived from *A. nitens,* its stratigraphic predecessor, but may have emerged from some form close to *A. simplicidens.*

Measurements* of Dentition of *A. reticulatus* (Type)
(UCMP 105039)
(decimal values in millimeters)

LP³	L^P	LM¹	WM¹	NPC¹	NMC¹	
1.0	2.8	1.9	2.8	5	5	

TC¹	TC²	PAPR¹	MLC¹	WM²	ACM^P	
22	16	2	0.15	2.7	0.15	

* See Appendix 1 for definitions of variates.

Allomys tessellatus,* new species
(Fig. 16; pl. 11a)

*Type.*Left mandible fragment with M₁₋₃, UCMP 105038.

Type locality. Haystack 19 (V–6587). See detailed locality description in Appendix 2.

Stratigraphic distribution. Single specimen only, from Haystack Valley Member of John Day Formation. Uppermost part of *Entoptychus-Gregorymys* Concurrent-range Zone. *Entoptychus individens* teilzone. Above teilzone of *A. reticulatus.*

Geographic distribution. Type is only known specimen. Drainage of John Day River, Wheeler County, Oregon.

Age. Latest Arikareen; early Miocene.

Diagnosis. Lower cheek teeth anteroposteriorly shorter than in other species but not narrower than in *A. reticulatus;* posterior lobe of M₃ shortened. Crests subequally prominent, short, curving, joining one another to form similarly sized fossettids. Number of closed fossettids greater than in other species; number of fossettids in M₂ (NF2) 12. Anteroconid crest on M₂ shorter than in other species. Fossettid between internal protoconid crest and anterolingual crest of mesoconid oval, shorter than in *A. reticulatus* and other species. Crest from entoconid reaching internal protoconid crest and dividing central fossettid, which is undivided in other species. Anterointernal crest of entoconid strongly joined to mesostylid. Occlusal surfaces of crests nearly approximating a single plane in earlier stage of wear than in other species. Lingual margin of M₂, M₃ more nearly straight anteroposteriorly than in other species (margin defined as line connecting metaconid, mesostylid, entoconid). Metastylid crest with pronounced inflection between metaconid and mesostylid, bounded ventrally by distinct lip.

Discussion. Like *A. reticulatus,* this species is characterized by small size and short, subequally developed crests that are rather evenly distributed throughout the occlusal area. The more frequent union of crests in *A. tessellatus* seems to represent a continuation of a trend toward more even partitioning of space commenced in *A. cavatus* and

---

*tessellatus; L. formed in mosaic, referring to pavementlike occlusal surface of small cells.

accentuated in *A. reticulatus,* and this could be interpreted as evidence for derivation of *A. tessellatus* from *A. reticulatus.* However, the anteroposterior compression of the cheek teeth, the shortening of the hypoconulid lobe on M₃, and the relatively narrow trigonid are unlike characters of any of the earlier species of the genus.

The rocks of the Haystack Valley Member in which *A. tessellatus* occurs represent the earliest occurrence of extensive sorted clastics in the formation and a transition to a new environment of sizeable streams and floodplains, suggesting that *A. tessellatus* may have been adapted to a different environment than that of its predecessors. Because stream environments must have existed outside the John Day basin during the earlier intervals, *A. tessellatus* may have immigrated from such a neighboring region. The relatively narrow trigonid and prominent hypoconid in *A. tessellatus* are curious because these are primitive conditions shared with taxa of prosciurine grade, such as *Plesispermophilus angustidens.* However, because this character is not shared by any of the other allomyines, even *Parallomys* or *A. cavatus,* it may be a late acquisition in the ancestry of *A. tessellatus.*

Measurements* of Dentition of *A. tessellatus* (Type)
(UCMP 105038)
(decimal values in millimeters)

LM1	LTR1	WTR1	NF2	LAC2
1.9	0.87	1.7	12	0.60

* See Appendix 1 for definitions of variates.

Alwoodia,[8] new genus

Type species. Alwoodia magna, n. sp.

Definition. Cheek teeth larger but lower incisor narrower than in *Allomys.* Length of M₁ (LM1) 2.5–3.0 mm, width of lower incisor (WI) usually 1.1 mm or less. Crests of cheek teeth higher, heavier, more elongate than in *Allomys* or *Parallomys,* lacking short accessory processes. Upper cheek teeth with elongate, excursive crests joining cusps of protoloph, metaloph; central transverse valley lacking crests accessory to protoloph, metaloph; central valley constricted by invading major crests of protoloph, metaloph. Labial faces of paracone, metacone more vertical than in *Parallomys.* Posterointernal crest of mesoconid connecting to hypoconid, enclosing posterolabial fossettid. Internal crests of lower molars less randomly oriented than in *Allomys* or *Parallomys,* trending dominantly posterolinguad.

Species. Alwoodia magna, late Olig., Ore.

Discussion. This group occurs lowest in the section of all allomyines in the stratigraphic collection, yet exhibits a morphology which, in crest height at least, is the most advanced and, unlike that of *Allomys,* had undergone enlargement of the primitive complement of crests as opposed to acquiring additional crests. Crests had become longer and thicker, rather than more numerous and thinner. The large size and absence of accessory crests are shared with the species *Allomys harkseni,* described by

---

[8] Honoring Dr. Albert E. Wood and his immense contribution to our knowledge of fossil rodents.

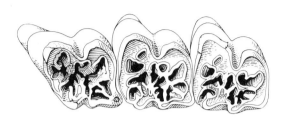

FIG. 16. *Allomys tessellatus,* left M_{1-3}, occlusal view, UCMP 105038, Haystack 19 (V–6587); labial top, X10.

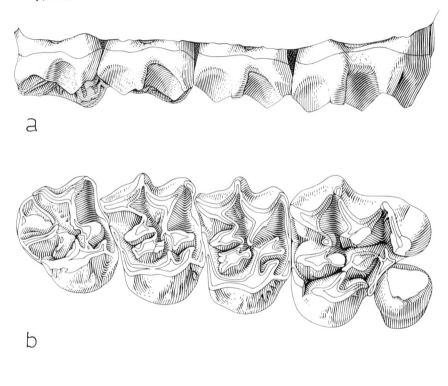

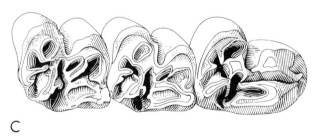

FIG. 17. *Alwoodia magna: a,* right P^3-M^3, labial view, UCMP 76941, Picture Gorge 22 (V–66116); *b,* same, occlusal view; *c,* left P_4-M_2, occlusal view, UCMP 76946, Picture Gorge 22; anterior right, X10.

J. R. Macdonald (1963:177) from the Monroe Creek Formation. The relationship of that species, which is now represented by numerous isolated teeth and a mandible (L. J. Macdonald, 1972; J. R. Macdonald, 1970), to the John Day allomyines is complicated by the possible presence there of more than one taxon and will be discussed in a separate study.

Alwoodia magna,[9] new species
(Fig. 17; pls. 11b-d, 12)

Type Right maxillary, P³-M³; tip of P³ missing; UCMP 76941.

Type locality. Picture Gorge 22 (V-66116); see detailed locality description in Appendix 2.

Stratigraphic distribution. Turtle Cove Member of John Day Formation. Lower part of *Meniscomys* Concurrent-range Zone. Teilzone of *Meniscomys uhtoffi* and slightly above. From about 5 m above Picture Gorge ignimbrite to stratum resting on Deep Creek tuff. Immediately beneath teilzone of *Allomys simplicidens*.

Distribution. Drainage of John Day River, Grant and Wheeler counties, Oregon.

Age. Early Arikareean; late Oligocene.

Referred specimens. Picture Gorge 12 (UWA 9591): UWBM 43036, mandible M₂₋₃; UWBM 43035, mandible M₁₋₂. Picture Gorge 22 (V-66111; UWA 5172): UCMP 76938, skull, DP³⁻⁴, M¹⁻³, missing tip of rostrum, zygomatic arches, posterior cranium; UCMP 76946, mandible, P₄-M₂; UCMP 76945, mandible, M₁₋₂; UCMP 105022, mandible, M₁₋₂; UCMP 76933, mandible, P₄-M₁; UCMP 105023, mandible, M₂₋₃; UWBM 39520, mandible, M₂₋₃. Picture Gorge 29, level 2, (UWA 9596), UWBM 47336, mandible, P₄-M₃. Haystack 32 (V-6581): UCMP 76995, M₂; UCMP 105021, M³. Picture Gorge 17 (UWA 5171): UWBM 43278, maxillary, M²⁻³.

Diagnosis. Only species of genus.

Measurements* of Dentition of *Alwoodia magna* (Type)
(UCMP 76941)
(decimal values in millimeters)

LP³	L^P	LM¹	WM¹	NPC¹	NMC¹	TC¹
1.5	3.6	2.4	3.6	0	2	8

TC²	PAPR¹	MLC¹	WM²	ACM^P
8	0.0	0.07	3.4	0.0

*See Appendix 1 for definitions of variates.

[9] *magna:* L. in reference to the size of the cheek teeth, which are larger than those of other allomyines in the John Day Formation.

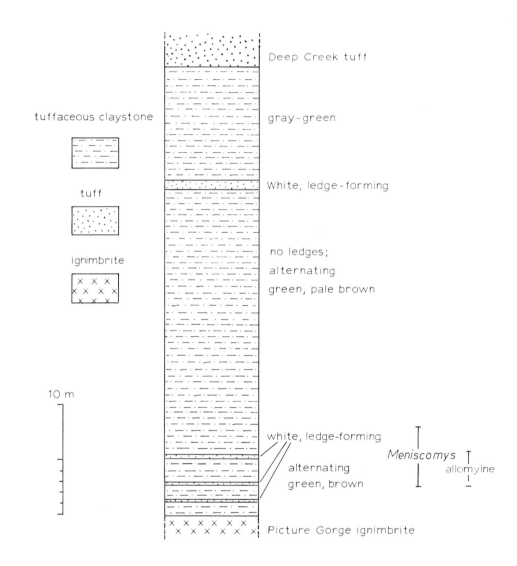

FIG. 18. Stratigraphic section at Picture Gorge 12 (V–6685, UWA 9591). Also see Appendix 2.

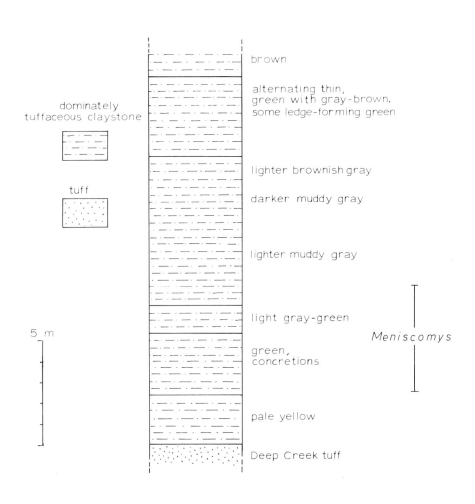

FIG. 19. Stratigraphic section at Picture Gorge 17 (UWA 5171). Also see Appendix 2.

Stratigraphic Relationships of
Meniscomyine Sites

Rodents closely related to *Meniscomys hippodus* are the most abundant fossil aplodontoids in the stratigraphic collections from the John Day Formation. Nevertheless, only one site, Picture Gorge 7 (V–6506, UWA 5183) yielded specimens through a very thick stratigraphic interval. Seven additional sites in that area (fig. 1) yielded specimens, but at most of these only the interval immediately above or below the Deep Creek tuff is exposed. *Meniscomys* was recovered from as wide a geographic distribution as were *Entoptychus* and *Pleurolicus,* from Haystack Valley north of Kimberly to the Camp Creek drainage more than 90 km southwest (fig. 1).

The distribution of cheek tooth dentitions of *Meniscomys* at Picture Gorge 7 extends vertically 24 m from the top of the Deep Creek tuff to the lower part of level 3 (Rensberger:1971, fig. 69). Some incisor fragments were found still higher (level 4). This range is extended downward about 37 m by occurrences beneath the Deep Creek tuff at sites Picture Gorge 12 (V–6685, UWA 9591, fig. 18) and Picture Gorge 22 (V–6616, UWA 5172). A single specimen was found 3 m beneath the Picture Gorge ignimbrite, clearly in situ, at locality Picture Gorge 20 (V–66114, UWA 4556), which extends the total range for the composite section to 76 m.

The occurrences of *Meniscomys* at most of the sites in the Picture Gorge quadrangle bear clear stratigraphic relationships to one another (fig. 20) because of the common presence of either the Deep Creek tuff, a thick and distinctive deposit of volcanic ash (see Fisher, 1962; 1963), or the Picture Gorge ignimbrite, which is even more distinctive (Fisher, 1966). It can be seen from figure 20 that the base of the *Entoptychus* teilzone overlies the *Pleurolicus* teilzone, which in turn overlies the top of the *Meniscomys* teilzone. The top of the *Meniscomys* Concurrent-range Zone (fig. 21) is defined by the top of the *Pleurolicus* teilzone (Fisher and Rensberger, 1972).

The correlative intervals immediately above the Deep Creek tuff (Picture Gorge sites 17 [UWA 5171] and 19 [V–66113, UWA 9592] and levels 1 and 2 at Picture Gorge 7) will be collectively referred to as interval D (table 1) in the analysis of the samples of *Meniscomys*. The uppermost levels (3,4 in table 1) of occurrence of *Meniscomys* at Picture Gorge 7 will be termed interval E; no other locality produced a meniscomyine from so high a stratigraphic position. The interval beneath and close to the Deep Creek tuff (Picture Gorge 22) will be called interval C. The interval above and close to the Picture Gorge ignimbrite (site Picture Gorge 12) and that beneath the ignimbrite (Picture Gorge 20) will be called, respectively, interval B and interval A.

The *Meniscomys* -bearing strata at Picture Gorge 29 lie well above the Picture Gorge

TABLE 1

Stratigraphic Intervals Containing *Meniscomys*

Interval	Localities*	Markers
		Pleurolicus Teilzone
E	Picture Gorge 7, levels 2+,3,4	
D	Picture Gorge 7, levels 1,2 Picture Gorge 17 Picture Gorge 19 Weaver's, level 2 Schrock's 1, level 0	
C	Picture Gorge 22	Deep Creek tuff
B	Picture Gorge 12	Picture Gorge ignimbrite
A	Picture Gorge 20, level 2	

*No superpositional relationship implied, except where a dashed line separates intervals. Levels refer to local stratigraphic levels at a single site, with level 2 stratigraphically superimposed on level 1, and so forth. Other sites produced *Meniscomys* but cannot be placed in this stratigraphic framework on the basis of the pre-existing physical and biostratigraphic evidence (fig. 20). Sites at which correlation was uncertain are omitted and were not used in the sample analysis. What is known of the stratigraphic relationships of these other sites is discussed on p. 84.

ignimbrite, but the Deep Creek tuff is not prominent if present, so the physical correlation of these occurrences with most of the others in this quadrangle is not certain.

Meniscomys was recovered at two localities in the Kimberly quadrangle. At one, Haystack 32 (V–6581), the occurrence is overlain by beds containing *Pleurolicus,* but other means of correlation are lacking. At the other site, Haystack 33, no means of physical correlation nor proximity to a recognized biostratigraphic unit was found.

In the Monument quadrangle, *Meniscomys* was recovered at only one locality, Hinton's (UWB-1039), where neither key beds nor other means of independent correlation are available.

In the region of Camp Creek, far to the south, two localities yielded *Meniscomys*. At Schrock's 1 (V–6351, UWC 56) *Meniscomys* was found a short distance above a prominent, coarse-grained tuff containing unaltered black shards of obsidian. A similar unit occurs in the region of the northern outcrops of the John Day Formation discussed above, but never in the same local section with the Deep Creek tuff, which is altered chiefly to the mineral clinoptilolite. Zeolitization generally took place after some

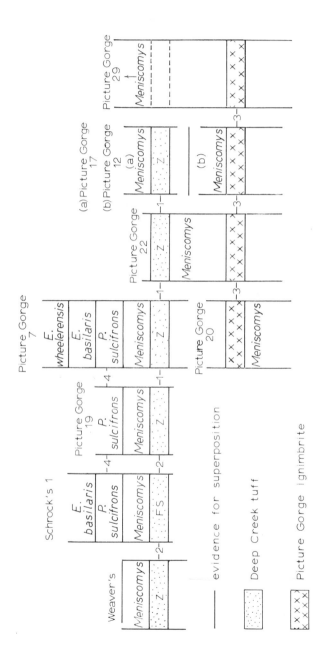

FIG. 20. Superpositional relationships of *Meniscomys* in the John Day Formation. Solid horizontal lines separating taxa or localities indicate physical evidence of superposition. Numerals 1, 2, 3 and 4 identify evidence for correlation based on (1) similar zeolitic (Z) facies of Deep Creek tuff, (2) similar positions of fresh obsidian facies (FS) and zeolitic facies of Deep Creek tuff in association with occurrences of *Meniscomys*, (3) unique lithology of the Picture Gorge ignimbrite, and (4) corresponding teilzones of *Pleurolicus sulcifrons* and *Entoptychus* (see Rensberger, 1971; 1973b). Physical identification of the Deep Creek tuff at Picture Gorge 29 is uncertain.

Vertical distances do not reflect rock thicknesses. Geographic positions of localities are described in Appendix 2.

Abbreviations: *E.* = *Entoptychus, P.* = *Pleurolicus.*

deformation had occurred; it therefore frequently provides different facies for rocks of the same age from locality to locality (Hay, 1963), raising the possibility that the two tuffs had a common origin and are of equivalent age (Fisher and Rensberger, 1972:15). A thick tuff resembling the Deep Creek tuff occurs a few kilometers east of Schrock's 1. This eastern locality, Weaver's (V–6645), produced a single secimen of *Meniscomys* from the beds just above the prominent tuff. These occurrences of *Meniscomys* in association with each of the two tuffs is the faunal evidence that led Fisher and Rensberger to believe that the zeolitic and obsidian tuffs represent the same ash fall. If, as Fisher and Rensberger concluded, the tuffs in the Camp Creek region are equivalent to the Deep Creek tuff in the Picture Gorge quadrangle, then the occurrences of *Meniscomys* at Schrock's 1 and at Weaver's represent interval D in the southern region.

Sample Relationships of Meniscomyines

Introduction

The relatively extensive vertical distribution of the fossils resembling *Meniscomys* suggests that the time range of the genus may have been sufficient for immigrant populations to appear or even for change within a single lineage to occur. The objective of this analysis was to recognize discrete populations by testing the hypothesis that all samples from the various stratigraphic levels and geographic sites are uniform in morphology. The results indicate that four groups, each distinguishable by a number of characters, are represented and that three of these groups were stratigraphically successive. Each of the successive groups is morphologically advanced over the preceding group, that is, each is distinguished by characteristics involving increased hypsodonty and lophodonty over the same variates in preceding groups.

Stratigraphic intervals. In order to simplify the stratigraphic relationships, localities and locality-levels that are known to be similar in age based on independent criteria are grouped as intervals (A, B, C, D and E, as summarized in table 1). Intervals, localities, or locality-levels were compared with one another with respect to each item of a list of morphologic characters (chosen as described in Methods of Analysis, above).

Criteria for differences. Significant differences in the frequencies of occurrence of character states at different intervals, localities or levels were considered indications of distinct populations at those positions. The frequency distribution for each character at each locality-level was also examined for evidence of polymodality, although at most positions the number of specimens was not sufficient to demonstrate complex modality.

Nature of the sample. Most specimens are mandibles or maxillaries. A total of 198 specimens of tooth-bearing jaws, skulls, or isolated teeth of meniscomyines were recovered from 19 different locality-levels, with distributions as follows:

Locality	Number of Specimens
Picture Gorge 7, level 1	28
Picture Gorge 7, level 2	72
Picture Gorge 7, level 3	6
Picture Gorge 7, level 4	3
Picture Gorge 12	2
Picture Gorge 17	32

Picture Gorge 19	2
Picture Gorge 20	1
Picture Gorge 22	4
Picture Gorge 29, level 2	1
Picture Gorge 29, level 3	1
Picture Gorge 29, level 4	1
Picture Gorge 33	1
Rudio Creek 3	1
Haystack 32	2
Haystack 33	25
Hinton's	1
Schrock's 1, level 0	13
Weaver's	2

Each specimen has been treated as a distinct individual, even though mandibular specimens are most often preserved as isolated halves, and some potentially belong to the same individual. The justification for this procedure is that the collections probably represent only small fractions of the total preserved parts of the populations, because (1) erosion soon destroys parts after they have weathered out, (2) the surface searching procedure for tiny elements in or on the clay mantle is inefficient, (3) most sites are on slopes with cracks or deep fissures which divert small elements during precipitation runoff, (4) most individual elements are widely scattered (those that are not are treated as a single individual), and (5) both elements of a pair, even if associated in the rocks, may not weather out simultaneously (see also Rensberger, 1971:13). In view of these conditions, recognition of each isolated dentition as a distinct individual seems reasonably accurate, whereas to consider only the minimum number of individuals (e.g., only right rami) would cause a loss of much pertinent information from collections that are small to begin with.

In the following subsections, the evidence for the distinctiveness of the three successive groups is presented, starting with that for the chronologically earliest. There follows a discussion of variation in the intermediate group, which is most abundant and varied. Finally, the samples with uncertain stratigraphic relationships are discussed, commencing with the relationships of a rather different lineage found at locality Rudio Creek 3. Specimens from the other localities of uncertain stratigraphic relationships are similar to those of the intermediate group, permitting correlation of those sites to the stratigraphic section. Names of dental structures are identified in figure 2, and the measured variates are defined in figure 3 and Appendix 1.

Primitive Group from Intervals A, B and C

Specimen from interval A. Only a single specimen is known from this interval. The specimen is, however, excellently preserved, consisting of an almost complete and undistorted skull with both mandibular rami. It is stratigraphically important because it is probably the oldest representative of *Meniscomys* known from the John Day Formation. It was found by Mr. Mike Uhtoff, who later accompanied me to the location at Picture Gorge 20 (UWA 4556). The stratigraphic position of the specimen is precisely

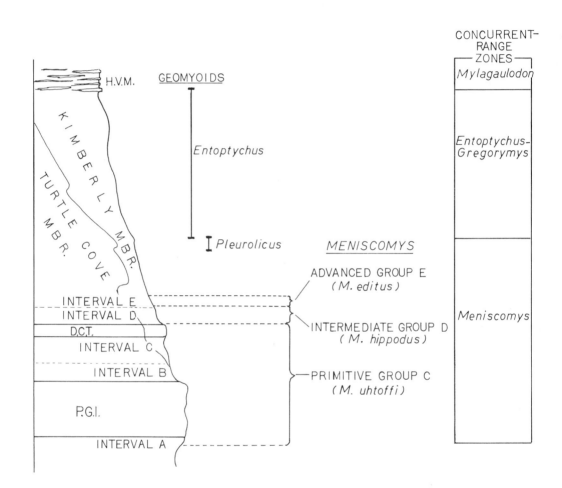

FIG. 21. Stratigraphic relationships of *Meniscomys* groups. Abbreviations of physical markers: P.G.i. = Picture Gorge ignimbrite; D.C.t. = Deep Creek tuff; H.V.M. = Haystack Valley Member of John Day Formation.

fixed: because of its position at the base of an overhanging welded tuff, there is no possibility that the specimen tumbled down from a higher position. The low stratigraphic position (fig. 21) of this occurrence, in interval A, is consistent with the morphology of the specimen.

The upper dentition of this specimen seems to be quite distinct from that of specimens of *Meniscomys* from stratigraphic intervals D and E (upper dentitions are not known from the superadjacent intervals B and C). The cheek teeth are well worn, and little of the morphology of the lower dentition can be accurately compared to that of the specimens occurring higher in the section. The upper dentition, on the other hand, does exhibit some comparable characteristics that suggest distinctiveness for this form.

The following list briefly describes the characters by which the upper dentition of the specimen from interval A differs from that of the specimens from interval D. The cited figures and tables give the quantitative relationships.

1. P^3 strongly grooved on posterior surface (HGP^3; fig. 23, table 2).
2. Anteroposterior diameter of P^3 (LP^3; fig. 24, table 2) greater.
3. Anteroposterior length M^3 (LM^3; fig. 22, table 2) greater.
4. Central fossettid on M^3 wider (WCF^3; fig. 22, table 2).
5. Presence of double mesostyle on M^3 (DMM^3; fig. 22, table 2).
6. Height of lingual enamel on P^4 (E^P; fig. 23).
7. Ratio of lingual length to transverse width of M^1 (LLM^1 to WM^1; fig. 38, table 2).

Because only a single specimen is known from interval A, the individual probabilities may not be regarded as significantly low. However, if no pair of these structures was influenced by the same developmental controls, then the probability of the aggregate probabilities occurring by chance is significantly low (less than 0.01). It is possible, however, that the characteristics involving greater anteroposterior length had a common genetic cause, leaving fewer independent groups of characters. However, the probability of the aggregate of the other tests is 0.02, which is still quite low.

In addition to the differences listed above, the height of enamel crown on P^3 (HP^3 in fig. 24) is much less than in any other specimen from the John Day Formation, and the height (0.08 mm) of lingual enamel on P^4 (E^P) is very close to the lower end of the range (0 mm) for the entire sample (fig. 23).

The M_1 is lingually relatively narrower than in specimens from interval E (see ratio of LLM^1 to WM^1 in fig. 38), suggesting a trend toward increasing lingual length of the tooth, perhaps from interval A upward.

Specimens from intervals B and C. Because the samples from intervals B and C are small and the specimens are morphologically too similar to be distinguished, these samples may be discussed as one. The increased size of the aggregate sample thus gained raises the significance of a number of tests.

As at the stratigraphic intervals immediately beneath and above the Picture Gorge ignimbrite, aplodontids are relatively rare in the deposits of interval C, which overlie those of interval B and extend up to the base of the Deep Creek tuff. Three mandibular fragments with moderately well preserved cheek dentitions, all from locality Picture

TABLE 2

Significance of Differences in HGP[3], LP[3], DMM[3], LM[3], WCF[3]
and Ratio of LLM[1] to WM[1] in *MENISCOMYS*
from Locality Picture Gorge 20 (Level 2)
and Interval D Localities.

Variate and Intervals (mm)*	Values Higher at†	Prob.‡	Locality-levels**
HGP[3] (prominent): (not prominent)	PG20(2)	.0909	PG20(2):Interval D
LP[3] (1.32–1.60):(>1.60)	PG20(2)	.0909	PG20(2):Interval D
DMM[3] (present):(absent)	PG20(2)	.0769	PG20(2):Interval D
LM[3] (1.44–1.76):(1.77–2.00)	PG20(2)	.0714	PG20(2):Interval D
WCF[3] (0.25–0.51):(0.52–0.66)	PG20(2)	.0909	PG20(2):Interval D
LLM[1] to WM[1] (0.43–0.48):(0.49–0.69)	Interval D	.0741	PG20(2):Interval D

*Except ratio.
†Locality with highest numerical values or where structure is present or more prominent or better developed.
‡Probability, in a 2 x 2 contingency table, of frequencies of values in morphological limits or character states listed in first column and locality-levels listed in fourth column, that no association between character state and locality exists.
**Locality-level or stratigraphic interval.

Abbreviations: PG7(1) = locality Picture Gorge 7, level 1; PG7(2) = Picture Gorge 7, level 2; PG12 = Picture Gorge 12; PG20(2) = Picture Gorge 20, level 2; PG 22 = Picture Gorge 22; PG17 = Picture Gorge 17; SK1(0) = Schrock's 1, level 0; PG7(3) = Picture Gorge 7, level 3. Localities and variates are defined in the appendices.

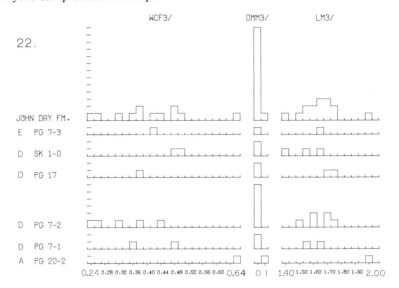

FIG. 22. Frequency distributions of width of central fossette on M³ (WCF³), presence of double mesostyle on M³ (DMM³), and length of M³ (LM³) in meniscomyines from different localities and stratigraphic levels of John Day Formation. Numerical values shown are basal for class. Locality abbreviations for this and subsequent frequency diagrams are: HS = Haystack; HI = Hinton; PG = Picture Gorge; RC = Rudio Creek; SK = Schrock's; ST = Stubblefield; WE = Weaver's. The two-part number indicates locality and level: for example, HS 8-1 = Haystack 8, level 1. Descriptions of localities are given in Appendix 2. Measurements (decimal values) are in millimeters. Designations A, B, C, D and E represent vertically successive stratigraphic intervals (see also table 1). Localities without these designations are of uncertain stratigraphic position.

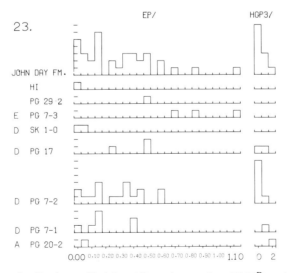

FIG. 23. Frequency distributions of height of lingual enamel on P⁴ (E^P) and prominence (on integer scale from 0 to 2) of posterior grooves on P³ (HGP³) in meniscomyines from different locality-levels in John Day Formation. See also legend of fig. 22. Measurements in millimeters.

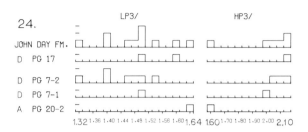

FIG. 24. Frequency distributions of length of P³ (LP³) and height of crown of P³ (HP³) in meniscomyines from different locality-levels of John Day Formation. See also legend of fig. 22. Measurements in millimeters.

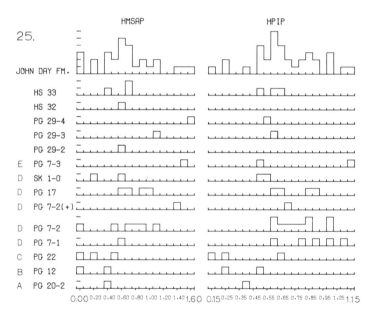

FIG. 25. Frequency distributions of mesostylid above anterolingual inflection on P₄ (HMSAP) and height of posterolingual inflection base on P₄ (HPIP) in meniscomyines from different locality-levels of John Day Formation. See also legend of fig. 22. PG 7-2(+) is stratigraphically higher than PG 7-2. Measurements in millimeters.

Gorge 22 (V–6616, UWA 5172), were recovered in this interval. This small sample shows no significant morphologic difference from that of interval B (Picture Gorge 12; V–6685, UWA 9591). It is possible that the group from interval C has a higher stylid on the anterior surface of the hypoconid on P_4 (HSHP in fig. 26 and table 4). This structure is higher in teeth from higher stratigraphic intervals, and no overlap in height exists between the specimens of intervals B and C. The significance test is indecisive due to the small sizes of the samples.

However, the specimens of intervals B and C taken together clearly differ in a number of characteristics from those of the overlying interval D:

1. Walls of labial inflection on P_4 less vertical (LIAP; fig. 29, table 3).
2. Mesostylid on P_4 less hypsodont (HMSP; fig. 30, table 3).
3. Mesostylid-metastylid notch on P_4 nearer base of crown (HMSAP; fig. 25, table 3).
4. Metaconid on P_4 less hypsodont (HMCP; fig. 29, table 3).
5. Base of posterolingual inflection on P_4 nearer base of crown (HPIP; fig. 25, table 3).
6. Stylid on P_4 hypoconid lower (HSHP; fig. 26, table 4).
7. Posterolingual inflection base on M_1 closer to base of crown (HPI1; fig. 28, table 5).
8. Anterolingual inflection base on M_1 closer to base of crown (HAI1; fig. 27, table 5).
9. Anterolingual inflection on P_4 transversely narrower (WAIP; fig. 26, table 3).
10. Anterior spur on hypoconulid of P_4 less frequently present (SHLP; fig. 30, table 4).
11. M_1 anteroposteriorly shorter (LM1; fig. 27, table 4).
12. Mesostylid process on P_4 absent (MP; fig. 31, table 4).
13. Talonid on M_2 transversely narrower (WT2; fig. 28, table 4).
14. Trigonid on M_2 anteroposteriorly shorter (LTR2; fig. 31, table 4).
15. Anteroposterior length of lower incisor in cross section less (LI; fig. 32, table 5).

These numerous differences indicate that the samples from intervals B and C together differ significantly from the samples derived from immediately overlying positions, those above the Deep Creek tuff (interval D). The characters by which the specimens from intervals B and C differ from those of interval D broadly fall into the functional categories of hypsodonty or lophodonty. Many of these characters involve either direct measurement of crown height or have been observed to accompany increases in crown height in other mammalian taxa.

In all characters in which a linear measure is involved, the values are smaller in the samples from the lower stratigraphic intervals. However, this does not mean that the specimens from the lower stratigraphic intervals differ in overall size. In other variates there are no significant differences in linear dimensions. For example, in the length of M_2, no appreciable difference in the mean exists between Picture Gorge 22 (2.06 ± 0.38 mm) and level 1 (2.00 ± 0.21 mm) or level 2 (1.98 ± 0.12 mm) of Picture Gorge 7, or Picture Gorge 17 (2.08 ± 0.47 mm). No significant difference in the length of P_4 was found. With the exception of the talonid on M_2, the samples showed no significant differences in overall width of the lower teeth.

TABLE 3

Significance of Differences in LIAP, HMSP,
HMSAP, HMCP, HPIP, and WAIP in *Meniscomys*
among Localities Picture Gorge 12,
Level 2 of Picture Gorge 20, Picture Gorge 22,
Levels 1 and 2 of Picture Gorge 7,
Picture Gorge 17 and Level 0 of Schrock's 1.

Variate and Intervals (mm)*	Values Higher at	Prob.	Locality-level
LIAP			
(0.04–0.11):(0.12–0.21)	PG22,PG12	.0079	PG12,PG22:PG17
(0.04–0.11):(0.12–0.21)	PG22,PG12	.0389	PG12,PG22:PG17, PG7(1,2)
(0.04–0.11):(0.12–0.21)	PG22,PG12	.1071	PG12,PG22:SK1(0)
(0.04–0.11):(0.12–0.21)	PG22,PG12, PG20(2)	.0238	PG20(2),PG12,PG22:PG17
HMSP			
(0.00–1.09):(1.10–1.49)	PG17	.0476	PG12,PG22:PG17
(0.00–1.09):(1.10–1.49)	PG17	.0333	PG20(2),PG12,PG22:PG17
(0.00–1.09):(1.10–1.49)	PG7(1,2)	.0924	PG20(2),PG12,PG22: PG7(1,2)
HMSAP			
(0.00–0.59):(0.60–1.19)	PG17,PG7(1,2)	.0429	PG12,PG22:PG17, PG7(1,2)
(0.00–0.59):(0.60–1.19)	PG7(1,2)	.0170	PG20(2),PG12,PG22: PG7(1,2)
(0.00–0.59):(0.60–1.19)	PG17,PG7(1)	.0041	PG20(2),PG12,PG22: PG17,PG7(1)
(0.00–0.59):(0.60–1.19)	PG17,PG7(1,2)	.0039	PG20(2),PG12,PG22: PG17,PG7(1,2)
HMCP			
(2.20–2.49):(2.50–3.19)	PG17,PG7(2)	.0357	PG12:PG17,PG7(2)
(2.20–2.49):(2.50–3.19)	SK1(0),PG17, PG7(2)	.0222	PG12:SK1(0),PG17,PG7(2)
HPIP			
(0.15–0.59):(0.60–1.19)	PG17,PG7(1,2)	.0005	PG12,PG22:PG17, PG7(1,2)
(0.15–0.59):(0.60–1.19)	PG7(1)	.0130	PG20(2),PG12,PG22: PG7(1)
(0.15–0.59):(0.60–1.19)	PG7(2)	.0010	PG20(2),PG12,PG22: PG7(2)
(0.15–0.59):(0.60–1.19)	PG17	.0238	PG20(2),PG12,PG22:PG17
(0.15–0.59):(0.60–1.19)	PG17,PG7(1,2)	.0001	PG20(2),PG12,PG22: PG17,PG7(1,2)
WAIP			
(0.20–0.34):(0.35–0.73)	PG7(1)	.0500	PG12,PG22:PG7(1)
(0.20–0.34):(0.35–0.73)	PG7(1,2)	.0179	PG12,PG22:PG7(1,2)

Note: For explanation of column headings see table 2.
*Except LIAP, which is measured in degrees.

TABLE 4

Significance of Differences in SHLP, LM1, MP, HSHP,
WT2, and LTR2 in *Meniscomys* from Localities Picture Gorge 12,
Level 2 of Picture Gorge 20, Picture Gorge 22,
Levels 1 and 2 of Picture Gorge 7, Picture Gorge 17 and
Level 0 of Schrock's 1.

Variate and Intervals (mm)	Values Higher at	Prob.	Locality-level
SHLP			
(absent):(present)	PG7(1,2)	.1032	PG12,PG22:PG7(1,2)
(absent):(present)	SK1(0),PG17, PG7(1,2)	.0357	PG12,22:SK1(0),PG17, PG7(1,2)
(absent):(present)	SK1(0)	.1429	PG12,PG22:SK1(0)
LM1			
(1.60–1.74):(1.75–1.84)	PG7(1)	.2000	PG12,PG22:PG7(1)
(1.60–1.79):(1.80–2.04)	PG7(2)	.0490	PG12,PG22:PG7(2)
(1.60–1.79):(1.80–2.04)	PG17	.0079	PG12,22:PG17
(1.60–1.79):(1.80–2.04)	PG17,PG7(1,2)	.0351	PG12,22:PG17,PG7(1,2)
MP			
(absent):(present)	PG7(1)	.0179	PG12,PG22:PG7(1)
(absent):(present)	PG7(2)	.0105	PG12,PG22:PG7(2)
(absent):(present)	PG17	.0040	PG12,PG22:PG17
(absent):(present)	PG17,PG7(1,2)	.0008	PG12,PG22:PG17, PG7(1,2)
(absent):(present)	PG7(1)	.0119	PG20,PG12,PG22:PG7(1)
(absent):(present)	PG17,PG7(1,2)	.0003	PG20,PG12,PG22:PG17, PG7(1,2)
HSHP			
(0.20–1.09):(1.10–2.29)	PG17,PG7(1,2)	.0256	PG12,PG22:PG17, PG7(1,2)
(0.00–1.09):(1.10–2.29)	PG17,PG7(1,2)	.0128	PG20,PG12,PG22:PG17, PG7(1,2)
(0.00–0.69):(0.70–1.09)	PG22	.1000	PG20,PG12:PG22
(0.00–0.19):(0.20–1.09)	PG12,PG22	.2000	PG20:PG12,PG22
WT2			
(1.35–1.54):(1.55–1.79)	PG17,PG7(1,2)	.0134	PG12,PG22:PG17, PG7(1,2)
LTR2			
(0.82–0.95):(0.96–1.19)	SK1(0),PG17, PG7(1,2)	.0079	PG12,PG22:SK1(0),PG17, PG7(1,2)

Note: For explanation of column headings see table 2.

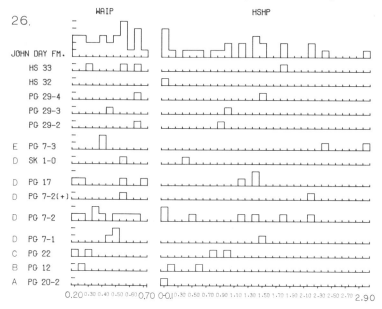

FIG. 26. Frequency distributions of width of anterolingual inflection on P_4 (WAIP) and height of anterior stylid on hypoconid of P_4 (HSHP) in meniscomyines from different locality-levels of John Day Formation. See also legend of fig. 22. Measurements in millimeters.

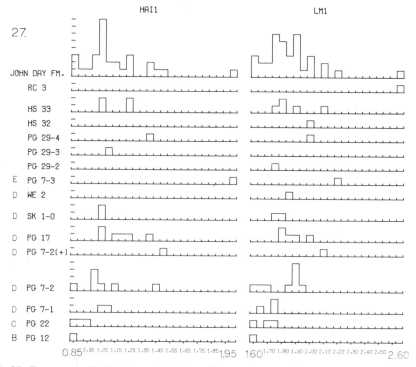

FIG. 27. Frequency distributions of height of base of anterolingual inflection on M_1 (HAI1) and length of M_1 (LM1) in meniscomyines from different locality-levels of John Day Formation. See also legend of fig. 22. Measurements in millimeters.

Instead, the differences involve characters contributing to hypsodonty or lopho-donty. The lingual inflection angle, LIAP, depends upon the parallelism of the sides of the crown of the tooth; this character commonly accompanies increased hypsodonty in mammals, usually during an early stage of the trend. Some characters are direct measures of the crown height of P_4 and M_1. Others involve lengths of specific occlusal crests; crest length (measured in any direction except parallel to occlusal motion) is a correlate of the rate of food subdivision (Rensberger, 1973a:520), and therefore may have been functionally related to a single lophodont trend. The increasing length of M_1 may be involved in the same trend. Molarization of premolariform teeth is a trend that often accompanies increases in lophodonty in herbivorous mammals (see Butler, 1952, and Savage, Russell, and Louis, 1965, for examples among the Equidae). In the specimens of *Meniscomys* from intervals A, B, and C, M_1 is samller than the adjacent cheek teeth. The subsequent increase in relative size of the tooth would have had an ef-fect on the cheek tooth series similar to that of molarization of premolars in other groups.

In view of these differences, the specimens from beneath the Deep Creek tuff must represent a group (or groups) distinct from and less advanced than those of interval D above this tuff. Few of the observed differences between groups C and D would preclude ancestry of group D from among group C. The frequency distributions show an overlap in many instances, and there are no new structures in group D that are not present in at least some individuals of group C, with the exception of the mesostylid process on P_4 (MP). Some of the specimens of group C occur close beneath the Deep Creek tuff, yet lack the process. The size of the mesostylid cusp alone is quite variable in the group from intervals B and C. Either variation is high or more than one at least partially isolated gene pool reached this area during interval C. In either case, some specimens with essentially the morphology expected in the ancestry of group D are pre-sent in the local underlying beds and in some cases are almost indistinguishable from those of interval D.

Intermediate and Advanced Groups from Intervals D and E

In the preceding section it was seen that the samples from the lower part of the sec-tion are morphologically distinct from those of the immediately higher stratigraphic in-terval D. The upper stratigraphic boundary of interval D most clearly divides the assemblage from the beds above interval C into morphologically distinguishable groups. The position of this boundary at Picture Gorge 7 is about 16 m above the Deep Creek tuff. The specimens from interval D differ from those of the overlying interval E in the following characteristics:

1. Base of anterolabial inflection on P^4 closer to base of crown (HALI^P; fig. 33, table 6).
2. Lingual enamel crown on P^4 lower (E^P; fig. 23, table 6).
   a. Ratio of height of lingual enamel on P^4 to length of M^3 (E^P to LM^3; fig. 33, table 6) smaller.
   b. Ratio of height of lingual enamel to length of P^4 (E^P to L^P; fig. 34, table 6) smaller even though P^4 shorter (L^P; fig. 36).

TABLE 5

Significance of Differences in HPI1, HAI1, and LI
in *Meniscomys* from Localities Picture Gorge 12,
Picture Gorge 20, Picture Gorge 22, Levels 1 and 2 of Picture Gorge 7,
Picture Gorge 17 and Level 0 of Schrock's 1.

Variate and Intervals (mm)	Values Higher at	Prob.	Locality-level
HPI1			
(0.30–0.69):(0.70–0.94)	PG7(1,2)	.1259	PG12,PG22:PG7(1,2)
(0.30–0.69):(0.70–0.94)	PG17	.0048	PG12,PG22:PG17
HAI1			
(0.85–0.99):(1.0–1.49)	PG7(1,2)	.0070	PG12,PG22:PG7(1,2)
(1.0–1.49):(0.85–0.99)	PG17	.0048	PG12,PG22:PG17
(0.85–0.99):(1.0–1.49)	SK1(0)	.0667	PG12,PG22:SK1(0)
LI			
(1.50–1.79):(1.80–2.19)	PG17,PG7(1,2)	.0376	PG12,PG22:PG17, PG7(1,2)
(1.50–1.79):(1.80–2.19)	PG17,PG7(1,2)	.0186	PG20,PG12,PG22:PG17, PG7(1,2)

Note: For explanation of column headings see table 2.

TABLE 6

Significance of Differences in HALI^P, E^P,
Ratio of E^P to LM^3, ratio of E^P to L^P, LL^P and CLP^P in *Meniscomys*
from Levels 1 and 2 of Picture Gorge 7, Picture Gorge 17,
Level 0 of Schrock's 1 and Level 3 of Picture Gorge 7.

Variate and Intervals (mm)*	Values Higher at	Prob.	Locality-levels
HALI^P			
(1.30–2.09):(2.10–2.59)	PG7(3)	.0221	PG7(1,2),PG17:PG7(3)
E^P			
(0.03–0.70:(0.71–1.15)	PG7(3)	.0003	PG7(1,2),PG17,SK1(0): PG7(3)
E^P to LM^3			
(0.10–0.29):(0.30–0.69)	PG7(3)	.1429	PG7(1,2),PG17,SK1(0): PG7(3)
E^P to L^P			
(0.02–0.15):(0.16–0.27)	PG7(3)	.0022	PG7(1,2),PG17:PG7(3)
LL^P			
(1.55–2.19):(2.20–2.44)	PG7(3)	.0015	PG7(1,2),PG17,SK1(0): PG7(3)
CLP^P			
(0.10–0.14):(0.15–0.44)	PG17,PG7(1,2), SK1(0)	.0008	PG7(1,2),PG17,SK1(0): PG7(3)

Note: For explanation of column headings see table 2.
*Except ratios.

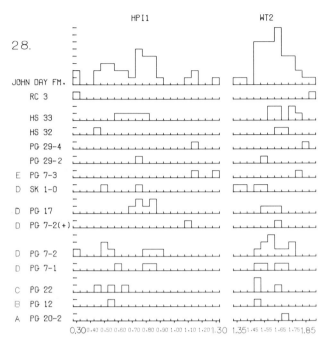

FIG. 28. Frequency distributions of height of base of posterolingual inflection on M_1 (HPI1) and width of talonid on M_2 (WT2) in meniscomyines from different locality-levels of John Day Formation. See also legend of fig. 22. Measurements in millimeters.

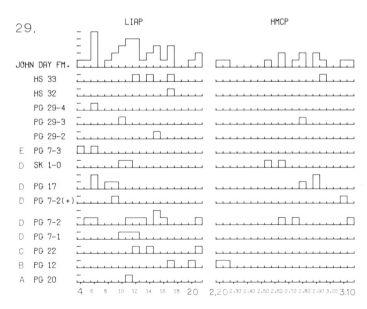

FIG. 29. Frequency distributions of vertical angle (degrees) between walls of labial inflection on P_4 (LIAP) and height of metaconid on P_4 (HMCP) in meniscomyines from different locality-levels of John Day Formation. See also legend of fig. 22. Decimal measurements in millimeters.

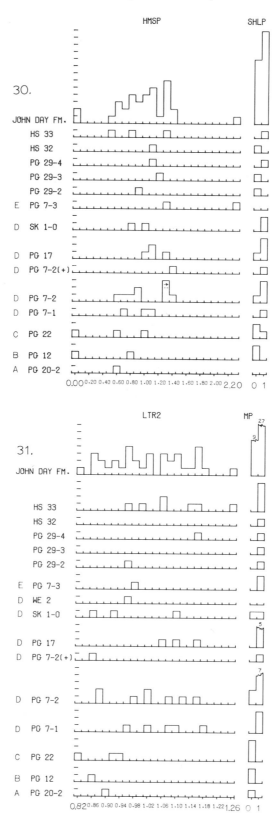

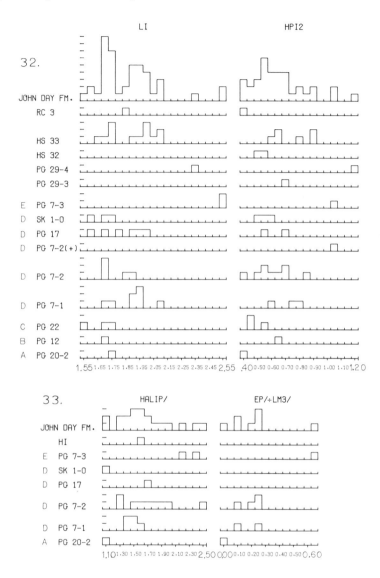

FIG. 30. Frequency distributions of height of mesostylid on P₄ (HMSP) and presence of anterior spur on hypoconulid of P₄ (SHLP) in meniscomyines from different locality-levels of John Day Formation. See also legend of fig. 22. Measurements in millimeters. Right arrow indicates a minimum value.

FIG. 31. Frequency distributions of length of trigonid on M_2 (LTR2) and presence of mesostylid process on mesoconid of P₄ (MP) in meniscomyines from different locality-levels of John Day Formation. See also legend of fig. 22. Measurements in millimeters.

FIG. 32. Frequency distributions of anteroposterior length of lower incisor cross section (LI) and height of posterolingual inflection base on M_2 (HPI2) in meniscomyines from different locality-levels of John Day Formation. See also legend of fig. 22. Measurements in millimeters.

FIG. 33. Frequency distributions of height of anterolabial inflection base on P⁴ (HALI[P]) and ratio of enamel height on P⁴ (E[P]) to length of M³ (LM³) in meniscomyines from different locality-levels of John Day Formation. See also legend of fig. 22. HALI[P] measured in millimeters.

3. Lingual moiety of P⁴ anteroposteriorly shorter (LLᴾ; fig. 34).
   a. Ratio of width to length of lingual loph of P⁴ greater (WLLᴾ to LLᴾ; fig. 35, table 7).
4. Paracone of P⁴ more convex lingually.
5. P⁴ shorter anteroposteriorly (Lᴾ; fig. 36, table 7).
6. Palate wider between P⁴'s.
7. Posterior segment of ectoloph on P⁴ anteroposteriorly shorter (LPEᴾ; fig. 36, table 7).
8. Parastyle and anterocone on P⁴ possibly narrower (WPLᴾ; fig. 37, table 7).
9. P⁴ transversely narrower (Wᴾ; fig. 37, table 7).
10. Posterior chevron on M¹ lower (CH¹; fig. 40, table 8), less well developed.
11. M¹ relatively shorter lingually (ratio of LLM¹ to WM¹; fig. 38, table 8).
12. M¹ anteroposteriorly shorter (LM¹; fig. 38, table 8).
13. M¹ transversely narrower (WM¹; fig. 39, table 8).
14. Walls of labial inflection of P₄ less parallel (LIAP; fig. 29, table 9).
15. Mesostylid of P₄ closer to base of posterolingual inflection (HMSP; fig. 30, table 9).
16. Mesostylid of P₄ closer to base of anterolingual inflection (HMSAP; fig. 25, table 9).
17. Stylid on hypoconid of P₄ shorter (HSHP; fig. 26, table 9).
18. Base of posterolingual inflection on M₁ closer to base of crown (HPI1; fig. 28, table 9).
19. M₁ anteroposteriorly shorter (LM1; fig. 27, table 10).
20. Base of anterolabial inflection on M₂ closer to base of crown (HAI2; table 10).
21. Base of posterolingual inflection on M₂ closer to base of crown (HPI2; fig. 32, table 10).
22. M₂ anteroposteriorly shorter (LM2; fig. 41, table 10).
23. Trigonid on M₂ possibly anteroposteriorly longer (LTR2; fig. 31, table 10).
24. Trigonid of M₂ relatively longer (ratio of LTR2 to LM2; fig. 41, table 10).
25. Talonid of M₂ possibly narrower (WT2; fig. 28, table 10).
26. Cross-sectional length of lower incisor less (LI; fig. 32, table 10).
27. Enamel of lower incisor thinner (TEI; fig. 42, table 10).

These differences indicate that the samples from stratigraphic interval D differ prominently from those of the overlying interval E (levels 2+ and 3 at Picture Gorge 7).

Size. The specimens of interval D are smaller than those of interval E. For example, measurements of anteroposterior length of P⁴ (Lᴾ, LPEᴾ) are about 15% larger at interval E.

Individuals from interval D seem to have had smaller skulls relative to cheek tooth size than those from E; the width between the P⁴ alveoli is 2.66 mm and 2.74 mm, respectively, in UW 31278 and 54900 from interval E, the mean of which is about 31% larger than the mean (2.06 mm) of 4 specimens (1.98 mm to 2.12 mm) from interval D at Picture Gorge 7. This difference exceeds the percentage difference in the premolar length of the same specimens (18.5%). The skull of UW 54900 is larger and more

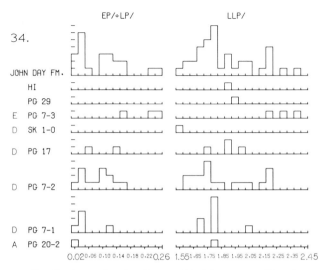

FIG. 34. Frequency distributions of ratio of enamel height on P⁴ (E^P) to length of P⁴ (L^P) and lingual length of P⁴ (LL^P) in meniscomyines from different locality-levels of John Day Formation. See also legend of fig. 22. LL^P measured in millimeters.

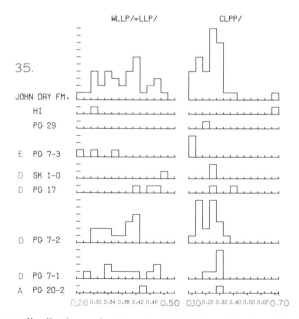

FIG. 35. Frequency distributions of ratio of width of lingual loph of P⁴ (WLL^P) to lingual length of P⁴ (LL^P) and depth of curvature of lingual surface of paracone on P⁴ (CLP^P) in meniscomyines from different locality-levels in John Day Formation. See also legend of fig. 22. CLP^P measured in millimeters.

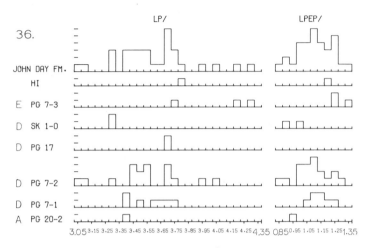

FIG. 36. Frequency distributions of length of P⁴ (L^P) and length of posterior segment of ectoloph on P⁴ in meniscomyines from different locality-levels of John Day Formation. See also legend of fig. 22. Measurements in millimeters.

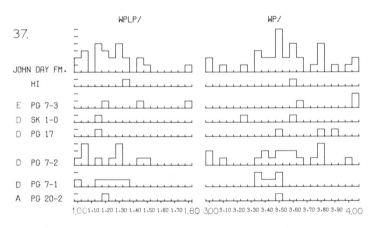

FIG. 37. Frequency distributions of width of parastyle and anterocone on P⁴ (WPL^P) and width of P⁴ (W^P) in meniscomyines from different locality-levels of John Day Formation. See also legend of fig. 22. Measurements in millimeters.

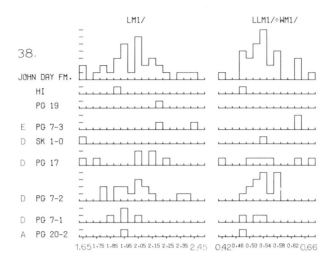

FIG. 38. Frequency distributions of length of M[1] (LM[1]) and ratio of lingual length of M[1] (LLM[1]) to width of M[1] (WM[1]) in meniscomyines from different locality-levels of John Day Formation. See also legend of fig. 22. Measurements of LM[1] in millimeters.

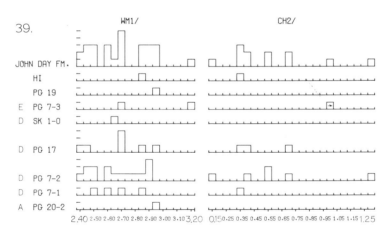

FIG. 39. Frequency distributions of width of M[1] (WM[1]) and height of enamel-dentine chevron on M[2] (CH[2]) in meniscomyines from different locality-levels of John Day Formation. See also legend of fig. 22. Measurements in millimeters.

TABLE 7

Significance of Differences in Ratio of WLL[P] to LL[P],
L[P], LPE[P], WPL[P] and W[P] in *Meniscomys* from
Levels 1 and 2 of Picture Gorge 7, Picture Gorge 17,
Level 0 of Schrock's 1 and Level 3 of Picture Gorge 7.

Variate and Intervals (mm)*	Values Higher at	Prob.	Locality-levels
WLL[P] to LL[P]			
(0.26–0.31):(0.32–0.52)	PG7(1,2),PG17, SK1(0)	.0565	PG7(1,2),PG17,SK1(0): PG7(3)
L[P]			
(3.05–4.19):(4.20–4.34)	PG7(3)	.0065	PG7(1,2),PG17,SK1(0): PG7(3)
(3.05–3.74):(3.75–4.34)	PG7(3)	.0078	PG7(1,2),PG17,SK1(0): PG7(3)
LPE[P]			
(0.85–1.24):(1.25–1.39)	PG7(3)	.0086	PG7(1,2),PG17,SK1(0): PG7(3)
WPL[P]			
(1.00–1.44):(1.45–1.84)	PG7(3)	.0727	PG7(1,2),PG17,SK1(0): PG7(3)
W[P]			
(3.00–3.64):(3.65–4.09)	PG7(3)	.0302	PG7(1,2),PG17,SK1(0): PG7(3)

Note: For explanation of column headings see table 2.
*Except ratios.

TABLE 8

Significance of Differences in CH[1], Ratio of LLM[1]
to WM[1], LM[1] and WM[1] in *Meniscomys*
from Levels 1 and 2 of Picture Gorge 7, Picture Gorge 17,
Level 0 of Schrock's 1 and Level 3 of Picture Gorge 7.

Variate and Intervals (mm)*	Values Higher at	Prob.	Locality-levels
CH[1]			
(0.35–1.39):(1.40–1.69)	PG7(3)	.1250	PG7(1,2),PG17,SK1(0): PG7(3)
LLM[1] to WM[1]			
(0.42–0.63):(0.64–0.69)	PG7(3)	.0085	PG7(1,2),PG17,SK1(0): PG7(3)
LM[1]			
(1.65–2.19):(2.20–2.49)	PG7(3)	.0246	PG7(1,2),PG17,SK1(0): PG7(3)
WM[1]			
(2.40–2.99):(3.00–3.24)	PG7(3)	.0690	PG7(1,2),PG17,SK1(0): PG7(3)

Note: For explanation of column headings see table 2.
*Except ratio.

TABLE 9

Significance of Differences in LIAP, HMSP, HMSAP, HSHP,
HPI1 and HAI1 in *Meniscomys* from Levels 1 and 2
(excluding 2 +) of Picture Gorge 7, Level 2 of Picture Gorge 7,
Picture Gorge 17, Level 0 of Schrock's 1, and Levels 2(+) and 3 of Picture Gorge 7.

Variate and Intervals (mm)*	Values Higher at	Prob.	Locality-levels
LIAP			
(4–7):(8–2l)	PG7(1,2),PG17, SK1(0)	.0789	PG7(1,2),PG17,SK1(0) :PG7(3)
(4–9):(10–21)	PG7(1,2),PG17, SK1(0)	.0632	PG7(1,2),PG17,SK1(0) :PG7(2 + ,3)
HMSP			
(0.00–1.29):(1.30–2.39)	PG7,PG7(2 + ,3)	.0421	PG7(1,2),PG17,SK1(0) :PG7(2 + ,3)
(0.00–1.29):(1.30–2.39)	PG7(2 + ,3)	.0095†	PG7(1,2),PG17,SK1(0) :PG7(2 + ,3)
HMSAP			
(0.00–1.19):(1.20–1.59)	PG7(2 + ,3)	.0158	PG7(1,2),PG17,SK1(0) :PG7(2 + ,3)
HSHP			
(0.00–2.29):(2.30–3.09)	PG7(3)	.0110	PG7(1,2),PG17,SK1(0) :PG7(3)
(0.00–2.19):(2.20–3.09)	PG7(2 + ,3)	.0088	PG7(1,2),PG17,SK1(0) :PG7(2 + ,3)
HPI1			
(0.30–0.94):(0.95–1.34)	PG7(2 + ,3)	.0083	PG7(2):PG7(2 + ,3)
(0.30–0.94):(0.95–1.34)	PG7(2 + ,3)	.0035	PG7(1,2):PG7(2 + ,3)
(0.30–0.94):(0.95–1.34)	PG7(2 + ,3)	.0119	PG17:PG7(2 + ,3)
(0.30–0.94):(0.95–1.34)	PG7(2 + ,3)	.0008	PG7(1,2),PG17,SK1(0) :PG7(2 + ,3)
HAI1			
(0.85–1.49):(1.50–2.04)	PG7(2 + ,3)	.0278	PG7(2):PG7(2 + ,3)
(0.85–1.49):(1.50–2.04)	PG7(2 + ,3)	.0182	PG7(1,2):PG7(2 + ,3)
(0.85–1.49):(1.50–2.04)	PG7(2 + ,3)	.0357	PG17:PG7(2 + ,3)
(0.85–1.49):(1.50–2.04)	PG7(2 + ,3)	.0058	PG7(1,2),PG17,SK1(0) :PG7(2 + ,3)

Note: For explanation of column headings see table 2.
*Except LIAP, which is measured in degrees.
†Individuals in which apex of mesostylid is incomplete due to wear have not been included in this calculation.

TABLE 10

Significance of Differences in LM1, HAI2, HPI2, LM2, LTR2,
Ratio of LTR2 to LM2, WT2, LI and TEI in *Meniscomys*
from Levels 1 and 2 of Picture Gorge 7, Level 2 of Picture Gorge 7,
Picture Gorge 17, Level 0 of Schrock's 1 and Levels 2(+) and 3 of Picture Gorge 7.

Variate and Intervals (mm)*	Values Higher at	Prob.	Locality-levels
LM1			
(1.60–2.04):(2.05–2.24)	PG7(2+,3)	.0182	PG7(2):PG7(2+,3)
(1.60–2.04):(2.05–2.24)	PG7(2+,3)	.0476	PG17:PG7(2+,3)
HAI2			
(0.40–1.34):(1.35–2.04)	PG7(2+,3)	.0735	PG7(1,2),PG17,SK1(0): PG7(2+,3)
(0.40–1.34):(1.35–2.04)	PG7(3)	.0625	PG7(1,2),PG17,SK1(0): PG7(3)
HPI2			
(0.40–0.89):(0.90–1.09)	PG7(2+,3)	.0110	PG7(1,2):PG7(2+,3)
LM2			
(1.75–2.19):(2.20–2.39)	PG7(2+,3)	.0222	PG7(2):PG7(2+,3)
(1.75–2.19):(2.20–2.39)	PG7(2+,3)	.0286	PG7(1,2),PG17,SK1(0): PG7(2+,3)
LTR2			
(0.86–0.99):(1.00–1.19)	PG7(1,2),PG17, SK1(0)	.1333	PG7(1,2),PG17,SK1(0): PG7(2+,3)
LTR2 to LM2			
(0.39–0.43):(0.44–0.65)	PG7(1,2),PG17, SK1(0)	.0058	PG7(1,2),PG17,SK1(0): PG7(2+,3)
WT2			
(1.35–1.64):(1.65–1.84)	PG7(2+,3)	.1423	PG7(1,2),PG17,SK1(0): PG7(2+,3)
LI			
(1.60–2.14):(2.15–2.59)	PG7(3)	.0476	PG7(2):PG7(3)
(1.60–2.14):(2.15–2.59)	PG7(3)	.0357	PG17:PG7(3)
(1.60–2.14):(2.15–2.59)	PG7(3)	.0036	PG7(1,2),PG17,SK1(0): PG7(3)
TEI			
(0.02–0.05):(0.06–0.08)	PG7(3)	.0095	PG7(1,2):PG7(3)
(0.02–0.05):(0.06–0.08)	PG7(3)	.0130	PG7(1,2),PG17,SK1(0): PG7(3)

Note: For explanation of column headings see table 2.
*Except ratio.

massive than that of UW 31451 from interval A, the only other relatively complete skull known.

Hypsodonty. Other differences involve increases in crown height in younger beds. Because of the smaller overall size at interval D than at interval E, a smaller crown height alone does not necessarily mean that the specimens from interval D were less hypsodont than the younger group. For example, changes in HALI^P and E^P are indicative of differences in hypsodonty unless they measure merely an overall increase in size.

However, ratios of anteroposterior length to crown height indicate that much of the increase in crown height at interval E is in excess of the increase in anteroposterior length of the teeth.

Although values for the length of P^4 are greater at interval E, the ratios (fig. 34) of enamel height (E^P) to length of P^4 (L^P) are also significantly greater at interval E than at interval D (table 6). The increase in enamel height of P^4 is therefore in excess of the overall increase in size of this tooth and does reflect an increase in hypsodonty. As another example, M^3 is smaller relative to the size of the other cheek teeth, especially P^4, at interval E than at interval D. Because hypsodonty is greater at interval E, the ratio of E^P to LM^3 (fig. 33) gives maximal separation of the specimens from these intervals.

The distance between the apex of the dentinal tract and the base of the enamel on M^1 (CH^1) is another direct measure of hypsodonty. In a specimen from interval E, UW 31276, the difference in development of the dentinal tract on M^1 is even greater than the measured variate CH^1 indicates. The lingual margin of the tract is vertical, and the tongue of lingual enamel covers a very limited extent of the posterior surface of the tooth. This specimen has a narrow P^4 and is lowest in stratigraphic position among the specimens from interval E. In contrast to that specimen, in UCMP 105085 from interval D, the dentinal tract gradually widens toward the roots, the lingual margin of the tract is not parallel to the lingual profile of the tooth, and the lingual enamel extends farther onto the posterior face of the tooth.

If dentinal tracts in aplodontids were not normally obscured by the adjacent teeth, I believe the overall difference in the morphology of the enamel-dentine borders of all the cheek teeth would be more conspicuous than differences in most of the characters cited above. Phyletic acquisition and elongation of dentinal tracts is known in other groups of rodents with evolving hypsodonty. In *Entoptychus* (Geomyoidea) it appears to have been a response to a coarser or more abrasive diet and the advantage of maintaining a flat occlusal surface (Rensberger, 1973a; 1975a:15).

Lophodonty. The specimens of interval D are less lophodont than those of interval E. Anteroposteriorly aligned structures perpendicular to the direction of occlusal motion tend to be less elongate in relation to their transverse widths or the width of the entire tooth. For example, the smaller ratios (fig. 35) of the width of the lingual loph of P^4 (WLL^P) to its length (LL^P) at level 3 of Picture Gorge 7 (table 7) suggests that the shorter lingual loph is not due to smaller overall size alone. Cusps tend to be less flattened transversely; for example, the lingual enamel of the paracone is more convex at interval D, making the cusp less elongate anteroposteriorly.

Transitional forms. The transition between intervals D and E is not as abrupt a change in morphology as might be inferred from the large list of differences. A few

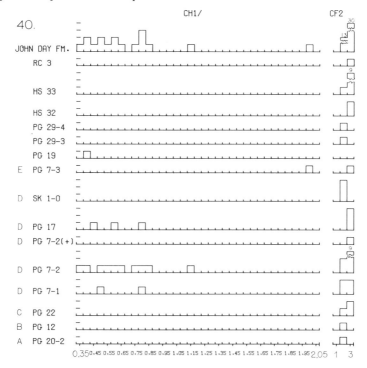

FIG. 40. Frequency distributions of height of enamel-dentine chevron on M^1 (CH^1) and degree of closure of central fossettid on M_2 (CF2) in meniscomyines from different locality-levels of John Day Formation. See also legend of fig. 22. Measurements of CH^1 in millimeters.

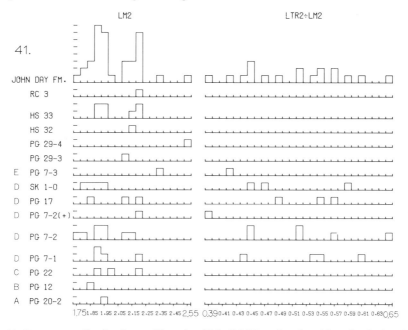

FIG. 41. Frequency distributions of length of M_2 (LM2) and ratio of length of trigonid on M_2 (LTR2) to length of M_2 in meniscomyines from different locality-levels of John Day Formation. See also legend of fig. 22. Measurements of LM2 in millimeters.

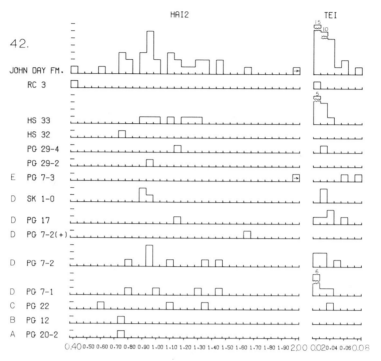

FIG. 42. Frequency distributions of height of anterolingual inflection base on M_2 (HAI2) and thickness of enamel on lower incisor (TEI) in meniscomyines from different locality-levels of John Day Formation. See also legend of fig. 22. Measurements in millimeters. Right arrow indicates minimum value.

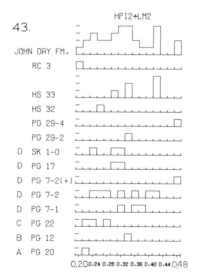

FIG. 43. Frequency distributions of ratio of height of posterolingual inflection base on M_2 (HPI2) to length of M_2 in meniscomyines from different locality-levels of John Day Formation. See also legend of fig. 22.

transitional forms found at the base of interval E and the top of interval D suggest that a group or groups of intermediate structure may have existed locally before the advanced members of group E appeared.

The width of M^1 (2.72 mm) in UW 31276, a specimen recovered from the base of interval E, is close to the mean value at each of the locality-levels of interval D (WM^1; fig. 39). UW 31276 is primitive in other variates as well ($HALI^P$ in fig. 33, LL^P in fig. 34, L^P in fig. 36, WPL^P in fig. 37, W^P in fig. 37, and LM^1 in fig. 38). Other variates argue against this specimen being a variant of group D, however. In several variates (ratio of E^P to L^P in fig. 34, CLP^P in fig. 35, LPE^P in fig. 36, and ratio of LLM^1 to WM^1 in fig. 38) the value for UW 31276 is comparable to that in one or more of the other specimens of interval E, and the ratio of WLL^P to LL^P (fig. 35) is slightly more advanced than the others. It seems likely that UW 31276 represents a distinct group of *Meniscomys* that is transitional in time and morphology between those of intervals D and E.

A similar relationship holds for a high specimen of interval D. The value of LIAP (fig. 29) in the single specimen (UW 39557) from the uppermost part (15 m position) of level 2 + at Picture Gorge 7 is larger than the mean at either of the lower levels of Picture Gorge 7 (interval D) but below the range of the specimens at Picture Gorge 7, level 3 (interval E). Values of this variate were shown to be diminishing prior to the time of deposition of interval D (table 3), and this trend appears to have continued to the transition of intervals D and E. The specimen is advanced in other variates as well.

Mosaic rates of change. Some variates changed upward through the section at a relatively steady pace. The length of M_1 increased between intervals C and D, and a change of similar magnitude appears to have occurred between intervals D and E. The value at interval E is only about 10% greater than the largest value at interval D.

Other variates changed more prominently at the transition between intervals D and E. For example, the height of the stylid on the hypoconid of P_4 (HSHP) was probably increasing from the lowest intervals (table 4) to the last known occurrences of *Meniscomys* in the formation, but the greatest change, about 36%, occurred between intervals D and E (fig. 26). Elevation of the anterolingual inflection on M_1 (HAI1) occurred between intervals C and D but became most conspicuous between intervals D and E (fig. 27). The value in a specimen from interval E is about 30% greater than the value in any specimen from interval D.

The trend toward increased diameter of the lower incisor (LI) begins at the transition between intervals C and D but has by far its greatest magnitude of change between intervals D and E (fig. 32). The values in the specimens from interval E are about 45% larger than the largest at interval D. The maximum thickness of the enamel on the incisor (TEI in fig. 42) is significantly smaller at interval D than at interval E (table 10), but there does not appear to be a significant change in the thickness of the enamel at the lower stratigraphic intervals.

The value of HAI2 in a specimen from interval E is about 38% larger than the largest from interval D (fig. 42). The value in the single specimen from the intermediate stratigraphic level, 2 + at Picture Gorge 7, falls among the high values of interval D, indicating that the more pronounced change occurred above level 2 +.

Diversity within Interval D

The large size of the sample from interval D (129 specimens) and the representation at a number of geographic localities (with one site, Schrock's 1, quite distant from the others) has provided evidence of geographic differences in morphology within interval D. Because some of the locality-levels that show such differences are very close geographically, probably fine stratigraphic differences are involved, suggesting that rapid shifts in distributions of subpopulations were occurring but were resolved somewhat differentially from locality to locality owing to preservational differences.

At the two most productive localities, Picture Gorge 7 and Picture Gorge 17, specimens were most common only a short distance above the Deep Creek tuff. Although specimens were found almost 18 m above the tuff at Picture Gorge 7, the abundance fell off above the 9 m level. The specimens from Picture Gorge 17 were confined to a vertical interval of about 5 m beginning a very short distance above the Deep Creek tuff.

Picture Gorge 7, levels 1 and 2; Picture Gorge 17. Each of the locality-levels within interval D, even those occurring as little as 3 km from one another, contain specimens with significant differences in morphology. The specimens from Picture Gorge 7, level 2, differ from those of level 1 in the following characteristics:

1. Enamel higher on anterior side of P^4 (HALI^P; fig. 33, table 11).
2. Lingual enamel of P^4 possibly more hypsodont (ratio of E^P to L^P; fig. 34, table 11).
3. Lingual surface of paracone on P^4 possibly less convex (CLP^P; fig. 35, table 11).
4. Lingual loph of P^4 transversely narrower (WLL^P; table 11).
5. M_1 and M_2 possibly longer anteroposteriorly (LM1, fig. 27; LM2, fig.41).

The samples from level 2, Picture Gorge 7 have the most widely ranging, most conspicuously polymodal distributions of values of any of the different locality-levels of interval D (HMSAP, fig. 25; HSHP, fig. 26; HPI1, fig. 28; HAI1, fig. 27; LTR2, fig. 31; ratio of E^P to L^P, fig. 34; LL^P, fig. 34; CLP^P, fig. 35; L^P, fig. 36; LPE^P fig. 36).

Yet these specimens represent more than a composite of the other locality-levels of this interval and include in some instances more primitive characters. In some variates the sample from level 2 at Picture Gorge 7 contains values more closely approaching those from intervals A, B and C than those of level 1 at this site. This is true of the height of the anterolingual inflection on M_1 (HAI1 in fig. 27), the height of the stylid on the hypoconid of P_4 (HSHP in fig. 26), the mesostylid process on P_4 (MP in fig. 31), and the width of the anterolingual inflection on P_4 (WAIP in fig. 26). A sample from Schrock's 1, an outlying locality discussed below, suggests one possible source of some of the heterogeneity found at level 2 of Picture Gorge 7.

The specimens from interval D at Picture Gorge 17 differ from those of Picture Gorge 7 (levels 1 and 2) in several characters:

1. Walls of labial inflection on P_4 more parallel than at levels 1 or 2 (LIAP; fig. 29, table 12).
2. M_1 anteroposteriorly longer (LM1; fig. 27, table 12) than at level 1.

TABLE 11

Significance of Differences in HALI^P,
Ratio of E^P to L^P, CLP^P, WLL^P and LM1
in *Meniscomys* from Levels 1 and 2 of Picture Gorge 7.

Variate and Intervals (mm)*	Values Higher at	Prob.	Locality-levels
HALI^P			
(1.10–1.69):(1.70–2.59)	PG7(2)	.0629	PG7(1):PG7(2)
E^P to L^P			
(0.02–0.05):(0.06–0.17)	PG7(2)	.0882	PG7(1):PG7(2)
CLP^P			
(0.10–0.29):(0.30–0.39)	PG7(1)	.0972	PG7(1):PG7(2)
WLL^P			
(0.52–0.71):(0.72–0.89)	PG7(1)	.0391	PG7(1):PG7(2)
LM1			
(1.60–1.79):(1.80–1.99)	PG7(2)	.0909	PG7(1):PG7(2)

Note: For explanation of column headings see table 2.
*Except ratio.

TABLE 12

Significance of Differences in W^P, LIAP and LM1 in *Meniscomys*
from Levels 1 and 2 of Picture Gorge 7 and Picture Gorge 17.

Variate and Intervals (mm)*	Values Higher at	Prob.	Locality-levels
W^P			
(3.00–3.79):(3.80–3.94)	PG17	.1278	PG17:PG7(2)
(3.00–3.79):(3.80–3.94)	PG17	.0833	PG17:PG7(1)
(3.00–3.79):(3.80–3.94)	PG17	.0786	PG17:PG7(1),PG7(2)
LIAP			
(4–9):(10–21)	PG7(2)	.0210	PG17:PG7(2)
(4–9):(10–21)	PG7(2)	.0286	PG17:PG7(1)
(4–9):(10–21)	PG7(1),PG7(2)	.0082	PG17:PG7(1),PG7(2)
LM1			
(1.60–1.79):(1.80–2.04)	PG17	.2308	PG17:PG7(2)
(1.60–1.79):(1.80–2.04)	PG17	.0179	PG17:PG7(1)
(1.60–1.79):(1.80–2.04)	PG17	.0747	PG17:PG7(1),PG7(2)

Note: For explanation of column headings see table 2.
*Except LIAP, which is measured in degrees.

These relationships suggest that the sample from Picture Gorge 17, though distinct, is related more closely to the specimens from level 2 than to those from level 1 at Picture Gorge 7 and that the sample from level 2 at Picture Gorge 7 may contain more than one local population, one close to that of the underlying level and another similar to that of Picture Gorge 17.

Schrock's 1. The specimens from Schrock's 1, a southernmost and rather distant site, differ in several characters from the specimens of interval D to the north along the John Day River:

    1. P^4 shorter anteroposteriorly (L^P; fig. 36, table 13).

    2. P^4 relatively shorter anteroposteriorly along lingual moiety (ratio of WLL^P to LL^P; fig. 35, table 13).

    3. M^1 shorter anteroposteriorly (LM^1; fig. 38, table 13).

    4. Enamel above anterolingual inflection on P^4 more restricted in depth ($HALI^P$; fig. 33, table 13).

    5. Central fossettid of M_2 not closed (CF2; fig. 40, table 13).

    6. Specimens probably smaller in many other variates.

Although distinct in these characters, the specimens from Schrock's 1 show a greater similarity to the specimens from interval D (north) than to those from intervals B and C. No specimens of meniscomyines were found above this occurrence at Schrock's 1, even though a thick section has produced other rodents, including specimens from the overlying *Pleurolicus* and *Entoptychus* teilzones.

Among the characters that differentiate specimens from interval D from those of intervals B and C, the specimens from Schrock's 1 in most instances are either intermediate or more closely resemble group D. The group from Schrock's 1 is intermediate in characters LIAP (fig. 29), MP (fig. 31), and HPIP (fig. 25); it more closely resembles group D in characters HAI1 (fig. 27), SHLP (fig. 30), HMSP (fig. 30), and WAIP (fig. 26). It more closely resembles the primitive group from intervals B and C in LTR2 (fig. 31, HMSAP (fig. 25), and LI (fig. 32); it resembles the specimens from intervals B and C in HSHP (fig. 26), but it also resembles the primitive specimens from level 2, Picture Gorge 7 in this character.

The specimens from Schrock's 1 therefore seem to represent a group distinct from any of those recovered from the region to the north but show characteristics of the groups from the intervals beneath and immediately above the Deep Creek tuff. The occurrence at Schrock's 1 is immediately above a prominent tuff resembling the black shard tuff (= Deep Creek tuff) in the northern region, which makes it contemporaneous with the early samples of interval D.

In conclusion, the samples from interval D are distinctive in the presence of considerable diversity, yet the lack of clear-cut trends suggests the existence of a widely distributed population that is represented locally by a series of subpopulations of short duration.

TABLE 13

Significance of Differences in L^P, Ratio of WLL^P to LL^P, LM^1, $HALI^P$, and CF2 in *Meniscomys* from Level 0 of Schrock's 1 and Northern Localities of Interval D.

Variate and Intervals (mm)*	Values Higher at	Prob.	Locality-levels
L^P (3.05–3.34):3.35–4.07)	North D	.0215	SK1(0):North D
WLL^P to LL^P (0.28–0.49):(>0.49)	SK1(0)	.0345	SK1(0):North D
LM^1 (<1.70):(1.70–2.39)	North D	.0741	SK1(0):North D
$HALI^P$ (<1.30):(1.30–2.09)	North D	.0667	SK1(0):North D
CF2 (complete):(incomplete)	North D	.0263	SK1(0):North D

Note: For explanation of column headings see table 2.
*Except ratio.
North D = Picture Gorge 7, levels 1 and 2, and Picture Gorge 17.

TABLE 14

Significance of Differences in LIAP, HMSP, HAI1, LM1, and LTR2 in *Meniscomys* from Haystack 33 and Localities of Intervals A (PG20), B (PG12, PG22), and D (PG17).

Variate and Intervals (mm)*	Values Higher at	Prob.	Locality-levels
LIAP (4–9):(10–21)	HS33	.0500	HS33:PG17
HMSP (0.50–0.99):(1.00–1.39)	PG17	.1428	HS33:PG17
HAI1 (0.85–0.99):(1.00–1.29)	HS33	.0143	HS33:PG12,22
LM1 (1.60–1.79):(1.80–1.99)	HS33	.0397	HS33:PG12,22
LTR2 (0.82–0.95):(0.96–1.27)	HS33	.0048	HS33:PG20,12,22

Note: For explanation of column headings see table 2.
*Except LIAP which is measured in degrees.

Locality-levels of Uncertain Stratigraphic Position

The preceding discussions considered only the locality-levels with physical evidence of stratigraphic position. Collections of *Meniscomys* and a related meniscomyine were obtained from some localities at which one or both of the key beds, the Deep Creek tuff and the Picture Gorge ignimbrite, are either missing, not exposed or not distinguishable. In the absence of either of these beds, it becomes impossible to establish the relative stratigraphic positions of samples from different geographic localities on the basis of criteria independent of the fauna itself. The remaining samples can placed only by morphologic comparisons of the fossils. For this purpose, the data on *Meniscomys* are the best available because the detailed biostratigraphic distributions published to date (Rensberger, 1971; 1973b) concern only taxa appearing stratigraphically above the known range of *Meniscomys*. First appearances of *Promerycochoerus* and *Paleocastor* occur at positions above the Deep Creek tuff, but these taxa are not common in this part of the section.

Locality Haystack 33 (V-66104). The fauna of Haystack 33 represents the intermediate group of *Meniscomys* from stratigraphic interval D, as shown by values of variates HMSAP (fig. 25), HAI1 (fig. 27), LM1 (fig. 27), MP (fig. 31), and LTR2 (fig. 31). The fauna less closely resembles that of Picture Gorge 17 than that of Picture Gorge 7 (e.g., LIAP, fig. 29) but tends to lack the wide variability present at level 2 of Picture Gorge 7. The most similar fauna is that from level 1 at Picture Gorge 7. The number of specimens recovered from this site is substantially greater than the number shown in the frequency distributions, for the teeth in many of the specimens were too worn to permit evaluation. Therefore the size of the collection from Haystack 33, which lies approximately 19 kilometers north of the localities in Picture Gorge quadrangle, supports the conclusion that interval D was a time of relative abundance of *Meniscomys*.

Locality Haystack 32 (V-6581). This site lies only 0.8 km to the northwest of Haystack 33 and is at approximately the same topographic elevation. The exposure consists of a small gulley (level 1) and a nearby steeper exposure representing a superimposed layer (level 2). Only two specimens of *Meniscomys* were recovered. One of these came from level 1. The other was collected in 1961, before I had commenced arbitrarily subdividing small exposures in the Turtle Cove Member. Most of the mammals were recovered from the gulley, and it is probable that both specimens of *Meniscomy* were found there. Level 2 yielded *Pleurolicus* (Rensberger, 1973:fig. 2) which has never been found in association with *Meniscomys* and consistently occurs stratigraphically above it.

The small sample from Haystack 32 seems more closely related to the primitive group from intervals B and C than to those from intervals D or E. The height of the anterolingual inflection base on M_2 (HAI2 in fig. 42) is less than in any specimen from intervals D or E. Values in three other variates, HSHP (fig. 26), LIAP (fig. 29), and HPI2 (fig. 32), suggest closest affinity to the primitive group.

Locality Picture Gorge 29 (V-6649, UWA 9596). The meniscomyines from this site are probably most closely related to the intermediate group of interval D. The occurrence of *Meniscomys* at Picture Gorge 29 is well above the Picture Gorge ignimbrite, but the Deep Creek tuff has not been identified at this site. The specimen from level 4

TABLE 15

Significance of Differences in HMSAP, WAIP, MP, HPI1, HAI1, HPIP and SHLP
in *Meniscomys* from Levels 2, 3, and 4 of Picture Gorge 29,
Level 2 of Picture Gorge 20, Picture Gorge 12, Picture Gorge 22,
Levels 1 and 2 of Picture Gorge 7 and Picture Gorge 17.

Variate and Intervals (mm)	Values Higher at	Prob.	Locality-levels
HMSAP (0.00–0.59):(0.60–1.69)	PG29(2,3,4)	.0048	PG29(2,3,4):PG20, PG12,PG22
WAIP (0.20–0.34):(0.35–0.69)	PG29(2,3,4)	.0500	PG29(2,3,4):PG12,PG22
MP (present):(absent)	PG29(2,3,4)	.0119	PG29(2,3,4):PG20, PG12,PG22
HPI1 (0.30–0.69):(0.70–1.19)	PG29(2,4)	.0667	PG29(2,4):PG12,PG22
HAI1 (0.85–0.99):(1.00–1.44)	PG29(3,4)	.0667	PG29(3,4):PG12,22
HPIP (0.15–0.54):(0.55–0.64)	PG29(3,4)	.1071	PG29(3,4):PG20,PG12,22
SHLP (present):(absent)	PG7(1,2),PG17	.2364	PG29(2,3,4):PG7(1,2), PG17

Note: For explanation of column headings see table 2.

occurred 2 m above a resistent tuff unit, the specimen from level 3 came from about 9 m below the tuff of level 4, and the specimen of level 2 came from a 1-m-thick ledge-forming tuff 3 m below the specimen of level 2. It is possible that one or the other of these units represents the Deep Creek tuff.

The specimens from the three levels of Picture Gorge 29 are advanced, without overlap of ranges, over those from intervals A, B and C in height of the mesostylid on P_4 (HMSAP in fig. 25), width of the anterolingual inflection on P_4 (WAIP in fig. 26) and absence of the mesostylid process on P_4 (MP in fig. 31). These specimens are also advanced in height of the posterolingual inflection base on M_1 (HPI1 in fig. 28) and height of the anterolingual inflection base on M_1 (HAI1 in fig. 27) over those of intervals B and C (table 15). The specimens of levels 3 and 4 are advanced in the height of the posterolingual inflection base on P_4 (HPIP in fig. 25) over five of six specimens from intervals A, B and C (table 15).

A spur is absent on the hypoconulid of P_4 (SHLP in fig. 30) in the specimens from levels 2 and 3, but it is present on the specimen from level 4 and in most specimens of interval D (table 15).

The evidence from most of these variates indicates that the specimens from all three levels of Picture Gorge 29 are advanced over those from intervals A, B and C, suggesting that the tuff just beneath level 2 may represent the Deep Creek tuff.

The specimen from level 4 may be as young as or younger than level 2 of Picture Gorge 7 and may come from a group unrepresented elsewhere. This specimen has a larger incisor cross section (LI in fig. 32) than any of the specimens of interval D. The width of the talonid on M_2 (WT2 in fig. 28) is greater than in any of the known specimens of intervals D and E.

Specimen from Rudio Creek 3 (V–66106). All three of the successive groups discussed above are referable to *Meniscomys*. At Rudio Creek 3, a form was found that represents a different type of meniscomyine. Only a single individual, consisting of a cranial fragment and mandible with partially complete M_{1-2} was recovered from this site, at which neither the Deep Creek tuff nor the Picture Gorge ignimbrite are present.

This specimen has several distinctive characteristics. The height of the posterolingual inflection base on M_1 (HPI1 in fig. 28) is less than that in any specimen from intervals B and C and all but one from interval D (UCMP 76803 of level 2, Picture Gorge 7). The first lower molar in the specimen from Rudio Creek 3 is, however, larger than that of UCMP 76803 or of any other specimen and the ratio of height of posterolingual inflection base to length of M_2 (HPI2 to LM2) shows that the inflection base is relatively lower than in any other meniscomyine from the formation (fig. 43).

Both M_1 and M_2 are larger than in other meniscomyines from the formation. For example, the width of the talonid on M_2 in the specimen from Rudio Creek 3 is greater than in any other specimen (fig. 28). The two molars seem to be essentially equal in size, whereas M_1 tends to be anteroposteriorly shorter than M_2 at other localities. The large size of the cheek teeth in the specimen from Rudio Creek 3 is not matched by a proportionately large incisor, however. The anteroposterior length of the incisor in cross section (LI) is near the mean for the other specimens and larger than in specimens from intervals A, B, and C (fig. 32).

The periotic is also distinctive. The septa, which are profuse and closely spaced in

Meniscomys, are fewer and more widely spaced. The tight, reticular structure present in *Meniscomys* is lacking.

Therefore the morphology is indicative of both primitiveness and advancement. Although very brachyodont, the specimen is advanced in cheek tooth size and closure of the central fossettid on M_2. In the morphology of the periotic, the specimen is more suggestive of the allomyines, but the cheek tooth pattern is that of a meniscomyine.

The thickness of the John Day Formation in this area is less than that of the other areas of occurrence of *Meniscomys* and the typically blue-green Turtle Cove Member, where it crops out in this region, is frequently pale red. The original topographic elevation of this region has been interpreted to have been greater than that of the areas to the west (center of Picture Gorge quadrangle) and northwest (Haystack Valley) because of the thinner deposits and evidence of oxidation (Fisher and Rensberger, 1972:19–21). The strata of Schrock's 1 are more similar in these characteristics to the deposits of Picture Gorge quadrangle. Consequently, none of the other areas of occurrence of *Meniscomys* may represent an environment like that at Rudio Creek 3. The presence of a species of *Allomys,* discussed in the following section, indicates that this site overlies the known stratigraphic range of *Meniscomys* at other sites and probably correlates with some part of level 4 at Picture Gorge 7. Some aplodontoidlike incisors were recovered from level 4 at Picture Gorge 7 but, although these seemed closest to *Meniscomys,* neither jaws nor cheek teeth of *Meniscomys* have yet been recovered from that level, and the incisors in some cases have features unlike those of *Meniscomys* from level 3. It is possible some of these represent *Rudiomys.*

Group Equivalence in Systematic Section

The groups of meniscomyines which differ prominently in morphology and stratigraphic or geographic separation are identified by the following taxonomic names in the section on systematics.

Group (Interval/Locality)	Taxon
E	*Meniscomys editus,* n. sp.
D	*Meniscomys hippodus* Cope
A,B,C	*Meniscomys uhtoffi,* n. sp.
Rudio Creek 3	*Rudiomys mcgrewi,* n. gen., n. sp.

Stratigraphic Relationships of Allomyine Sites

Rodents related to *Allomys nitens* Marsh, although not as abundant in collections as *Meniscomys,* are as widespread geographically and are distributed through an even greater stratigraphic interval. Allomyine rodents in the UCMP and UWBM collections range in the John Day Formation from Haystack Valley in the north to the Camp Creek region 90 km or so to the southwest (fig. 1). The stratigraphic range extends from the *Entoptychus individens* teilzone (fig. 44) at the top of the *Entoptychus-Gregorymys* Concurrent-range Zone (fig. 45) in the Haystack Valley Member, down to the base of the *Meniscomys* Concurrent-range Zone as defined by Fisher and Rensberger (1972:fig. 6). Although allomyines have not been recovered as low as the base of the *Meniscomys* teilzone, close beneath the Picture Gorge ignimbrite as now revised, few specimens of rodents of any kind have been recovered from below the ignimbrite.

The lowest occurrence of allomyines in the stratigraphic collections is that of several specimens at locality Picture Gorge 12 (UWA 9591), associated with *Meniscomys* a short distance above the Picture Gorge ignimbrite (fig. 20) and beneath the Deep Creek tuff. This position is stratigraphic interval B (table 16). A similar association was found close beneath the Deep Creek tuff (fig. 20) at locality Picture Gorge 22 (V-6616, UWA 5172), which represents interval C. A specimen from Picture Gorge 29, level 2 (UWA 9596) turns out in the following analysis to be related to the allomyine specimens from interval C. This specimen occurred on a tuff bed that apparently represents the Deep Creek tuff (p. 101). Consequently, for the purposes of this analysis, interval C is regarded as extending to the upper surface of the Deep Creek tuff (table 16).

Allomys is also associated with occurrences of *Meniscomys* in interval D, above the Deep Creek tuff (figs. 20, 44, and table 16) at Picture Gorge 7, levels 1 and 2 (V-6506, UWA 5183), Picture Gorge 17, (UWA 5171), Picture Gorge 19 (UWA 9592), Picture Gorge 29, level 3 (UWA 9596) and Schrock's 1, level 2 (V-6351). Although locality Haystack 33 (V-66104) can not be correlated with other sites on a physical basis, as was shown in the preceding section, it contains *Meniscomys hippodus* and therefore also correlates with the beds above the Deep Creek tuff (fig. 44).

Allomys was also found at interval E, above the occurrence of *Meniscomys hippodus* (levels 3 and 4 of Picture Gorge 7, level 1 of Haystack 6 [UWA 4799], level 4 of Schrock's 1, and at Picture Gorge 33 [V-66123, UWA 5833]). At each of these sites (fig. 44) the occurrence is beneath the *Pleurolicus* teilzone. Schrock's 1, level 8 (V-6351) yielded *Allomys* and *Pleurolicus* but no *Meniscomys,* an association

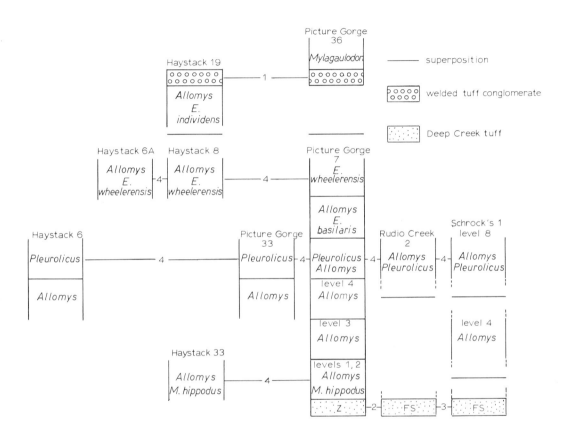

FIG. 44. Superpositional relationships of *Allomys* in the John Day Formation. Solid horizontal lines separating taxa or localities indicate physical evidence for superposition. Numbers indicate a correlation based on (1) conglomerate of welded tuff pebbles in lower part of Haystack Valley Member; (2) Deep Creek tuff, one facies of which has distinctive fresh black glass lapilli (FS), the other zeolitic glass (Z); (3) identical facies of the Deep Creek tuff; (4) corresponding teilzones of short lived species, *Meniscomys hippodus, Pleurolicus sulcifrons, Entoptychus basilaris,* and *Entoptychus wheelerensis,* of which the superpositional relationships are known (see the previous section of this paper and Rensberger, 1971; 1973b).

For superpositional relationships of allomyines at localities Picture Gorge 12, Picture Gorge 22, Picture Gorge 17, Picture Gorge 19 and Picture Gorge 29, see fig. 20. Allomyines are associated with *Meniscomys* at each of those sites. Vertical distances do not reflect rock thicknesses. Geographic positions of localities are described in Appendix 2.

represented also at Rudio Creek 2, level 1 (V–7077) and level 5 at Picture Gorge 7. *Pleurolicus* occurs in a restricted interval of the same age throughout the John Day Formation and above the *Meniscomys* teilzone (Rensberger, 1973b:8, fig. 5). The locality-levels within the *Pleurolicus* teilzone will be collectively referred to as interval E2, and the lower part of interval E, between the *Pleurolicus* and *M. hippodus* teilzones, as E1 (table 16).

The beds above the *Pleurolicus* teilzone (fig. 44) contain an abundance of *Entoptychus* (Rensberger, 1971:fig. 71) occasionally associated with *Allomys*. At eight of these sites the association is with primitive to intermediately advanced species of *Entoptychus (E. basilaris* through *E. cavifrons),* that occur stratigraphically low in the *Entoptychus* teilzone. These occurrences (level 6 at Picture Gorge 7 [V–6506], levels 1 and 2 at Haystack 8 [V–6322], Haystack 6A [V–6505], Haystack 1 [V–6429], and Rudio Creek 4 [V–6600, UWA 5929], level 10 at Schrock's 1 [V–6351] and Stubblefield 1A [V–6658]) will be referred to as interval F in the sample analysis (table 16).

The stratigraphically highest known occurrence of *Allomys* in the John Day Formation is in the Haystack Valley Member at Haystack 19 (V–6587). This stratigraphic position is identifiable because it is characterized by abundant sandstones that indicate a change in the depositional regime during tectonic activity late in John Day time (Fisher and Rensberger, 1972:21) and by the association of *Entoptychus individens* (Rensberger, 1971:88), the most advanced member of *Entoptychus* known (fig. 44). This position will be called interval G in the sample analysis (table 16).

The occurrences listed above form a stratigraphically and biostratigraphically controlled basis for studying samples taken from positions of similar age and for making comparisons between morphologic groups of different ages.

Allomyines were also found at four localities with neither physical nor independent biological associations that might identify the stratigraphic position. At one of these localities, Rudio Creek 3 (V–66106), an allomyine was found at the base of the exposure at about the same position as *Rudiomys mcgrewi,* a meniscomyine representing a lineage distinct from that of *Meniscomys.* No stratigraphically critical taxon was recovered from Picture Gorge 34 (UWA 5834) or Picture Gorge 43 (UWA 9966). The occurrence of *Allomys* at each of these sites is above the Picture Gorge ignimbrite and therefore above interval A. No specimens, even incisors, of *Entoptychus* were found, suggesting a position beneath the *Entoptychus* teilzone. For each of these sites, a position ranging anywhere from interval B to intervals E1 or E2 is possible (table 16). A specimen of *Allomys* was found at Picture Gorge 6 (UWA 9591), where the stratigraphic position is also uncertain.

TABLE 16
Stratigraphic Intervals Containing Allomyines

Interval	Localities*	Markers
G	Haystack 19	*E. individens* teilzone, Haystack Valley Member
F	Rudio Creek 4 Haystack 1 Haystack 8, levels 1,2 Haystack 6A Stubblefield 1A Picture Gorge 7, level 6 Schrock's 1, level 10	*E. cavifrons* teilzone to *E. basilaris* teilzone
E2	Schrock's 1, level 8 Rudio Creek 2, level 1	*Pleurolicus* teilzone
E1	Picture Gorge 33 Haystack 6, level 1 Schrock's 1, level 4 Picture Gorge 7, level 4 Picture Gorge 7, level 3	
D	Haystack 33 Picture Gorge 19 Picture Gorge 17 Schrock's 1, level 2 Picture Gorge 7, level 2 Picture Gorge 7, level 1 Picture Gorge 29, level 3	*M. hippodus* teilzone
C	Picture Gorge 29, level 2	Deep Creek tuff
	Picture Gorge 22 Haystack 32	
B	Picture Gorge 12	
		Picture Gorge ignimbrite

*No superpositional relationships implied, except where a dashed line separates intervals.

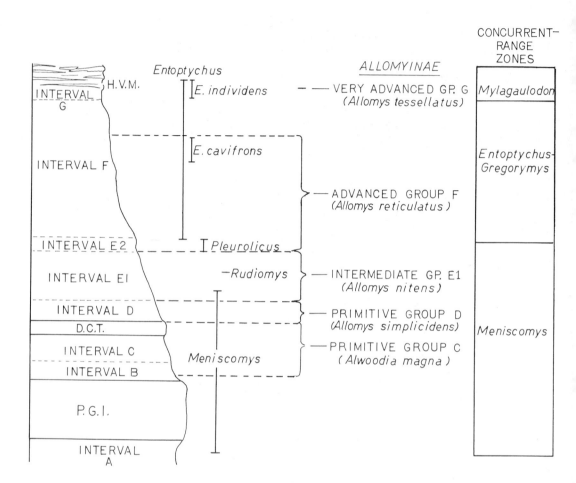

FIG. 45. Stratigraphic relationships of allomyine groups. See also legend of fig. 21.

Sample Relationships of Allomyines

General Characteristics

Aplodontid rodents resembling *Allomys* have one of the longest vertical distributions in the John Day Formation. The sample of dentulous jaws, teeth and skulls of allomyines consists of 80 specimens from 31 locality-levels, distributed as follows:

Locality	Number of Specimens
Picture Gorge 6	1
Picture Gorge 7, level 1	5
Picture Gorge 7, level 2	3
Picture Gorge 7, level 3	5
Picture Gorge 7, level 4	1
Picture Gorge 7, level 6	1
Picture Gorge 12	2
Picture Gorge 17	2
Picture Gorge 19	4
Picture Gorge 22	8
Picture Gorge 29, level 2	1
Picture Gorge 29, level 3	1
Picture Gorge 33	10
Picture Gorge 34	2
Picture Gorge 43	1
Stubblefield 1	1
Rudio Creek 2	1
Rudio Creek 3	2
Rudio Creek 4	2
Haystack 1	1
Haystack 6	6
Haystack 6A	1
Haystack 8	1
Haystack 8, level 1	1
Haystack 8, level 2	2
Haystack 19	1
Haystack 32	2
Haystack 33	1
Schrock's 1, level 4	1
Schrock's 1, level 8	9
Schrock's 1, level 10	1

The maximum local abundance of allomyines is less than that of *Entoptychus, Pleurolicus,* or *Meniscomys,* an observation that is not a result of the collecting technique because the number of localities is greater than those yielding meniscomyines and geomyoids, and an association of allomyines with the other genera is common. Whereas *Meniscomys* shows an increase in abundance at one stratigraphic interval (D), there is no comparable change in abundance of allomyines through their much longer stratigraphic range. For example, at intervals B and C, where *Meniscomys* is less abundant than in later deposits, an allomyine is more abundant than *Meniscomys* and is represented in the stratigraphic collections by 10 mandibles. At interval D, where 57 mandibles of *Meniscomys* were recovered, 10 lower dentitions of *Allomys* were found. At interval E2 *Meniscomys* has disappeared entirely but the allomyines are still represented by 7 occurrences of lower teeth. At intervals F and G, where *Entoptychus* is known from more than 1000 jaws, the allomyines are represented by 6 mandibles.

The consistently small size of the stratigraphically separated samples of allomyines makes recognizing morphologic changes more difficult than with the other taxa. A second difficulty is the absence of a pronounced and sustained trend, such as the increasing hypsodonty that characterized the histories of both *Entoptychus* and *Meniscomys.* The terminal population of the allomyines is almost as brachyodont as the initial group. On the other hand, the fact that the allomyines remained brachyodont is associated with a lesser absolute degree of wear on the cheek tooth crowns, so that cusp and crest structures are more commonly preserved. In addition, the occlusal structure in the allomyines (fig. 13) is exceedingly complex, equalling or surpassing that in the modern petauristine sciurids, and provides an abundance of visible morphologic characters potentially subject to differentiation.

As in other rodents, recovery and a preservational bias against upper dentitions and for lower dentitions exists, presumably resulting from a greater tendency for the maxillary rather than the dentary to break. The larger representation of mandibles has often made it easier to use them to recognize differences between samples.

The evidence indicates that the allomyine specimens represent at least five groups, each having distinctive morphologic characteristics and a generally distinct stratigraphic range. The morphologic evidence for the distinctiveness of each of these groups is presented below, in order of chronologic appearance. The earliest group, occurring at intervals B and C, represents a lineage that does not reappear after interval C. It was replaced by a group with smaller, less advanced dentitions near the base of interval D. These small forms were replaced by a group with larger cheek teeth and more complexly branching occlusal crests at interval E1. The group of interval E1 was replaced by forms with even more complexly crested but smaller teeth in interval E2. This group ranged upward through interval F, that is, through most of the *Entoptychus* teilzone. In interval G a small form appears with many of the tiny crests anastomosing to enclose occlusal lakes.

The following text summarizes the support for these conclusions. The evidence for the relationships between samples is based on material from those localities where specimen-yielding levels can be related to one another by means of superposition and key beds but also utilizes biostratigraphic data, including that provided by the analysis of the meniscomyine samples (see preceding section on stratigraphic relationships of allomyine localities). The names of dental structures are identified in figures 2 and 13, and the measured variates are defined in figure 13 and Appendix 1.

Group from Intervals B and C

Large cheek tooth size. The specimens from these intervals have larger cheek teeth than those from higher stratigraphic positions, as shown by the following characteristics:

1. Length of M_1 (LM1; fig. 46, table 17).
2. Width of talonid on M_1 (WTR1; fig. 46, table 17).
3. Length of P^3 (LP^3; fig. 51, table 18).
4. Length of M^1 (LM^1; fig. 51, table 18).
5. Width of M^1 (WM^1; fig. 51, table 18).

Among the specimens from higher positions, those from interval E1 most closely approach the size of specimens from intervals B and C.

In contrast to the larger cheek teeth, the lower incisor is significantly narrower than at higher stratigraphic intervals (WI in table 17 and fig. 48). This is true of both young and old individuals.

Crest simplicity and prominence. The crests of the cheek teeth at intervals B and C are more prominent and have a more directed, less divergent orientation than at higher intervals. Many of the crests of the lower cheek teeth are longer. For example, the anteroconid crest on M_2 is longer than at intervals D through E2 (LAC2 in table 17 and fig. 47).

It is difficult to obtain exact quantitative measures for differences in crest morphology because of variations in termination and branching, in union with other structures and in wear. Nevertheless, one can see that the crests in specimens from intervals B and C are straighter, more nearly parallel, heavier (pl. 11b) and lack minor processes characteristic of higher forms (pl. 10a-d). The crests are clearly higher than those at interval D (HC in table 17 and fig. 49).

The interiorly directed crest (IPC in fig. 13) from the protoconid of M_2 (this crest is always poorly developed on M_1) is almost straight in its course toward the anterior margin of the entoconid (fig. 17c), which brings it into union with the posterior end of the anteroconid crest. In specimens from intervals E1, E2, F and G the interior protoconid crest is directed less strongly posteriad, more toward the mesostylid than entoconid (pl. 9c), or it tends to loop posterolabiad and unite with the anterolingual process from the mesoconid (pl. 10b-d). The posterolingual-anterolabial alignment of the internal protoconid crest at intervals B and C is more or less paralleled by the attitudes of other internal crests except the hypolophid at its emanation from the hypoconid. Because of this common trend in the attitudes of the crests, the enclosed fossettids and valleys tend to be more elongate and subparallel than in specimens at intervals E1, E2, F and G, where shorter fossettids and divergent attitudes are dominant.

The marginal metastylid crest also tends, especially in M_2 and M_3, to be directed more strongly linguad (pl. 11b) than in specimens (e.g., pl. 10d) from intervals D, E1, E2, F or G. This attitude is associated with a more pronounced lingual convexity of the tooth margin, strongest in M_{2-3}, and greater lingual prominence of the mesostylid. The

TABLE 17
Significance of Differences in LM1, WTR1, WI, LAC2 and HC in Allomyines from Intervals B, C, D, E1 and E2.

Variate and Intervals (mm)	Values Higher at	Prob.	Stratigraphic Intervals
LM1			
(1.90–2.39):(2.40–2.99)	B,C	.0099	B,C:D,E1,E2
(1.90–2.39):(2.40–2.99)	B,C	.0539	B,C:D,E1
(1.90–2.39):(2.40–2.99)	B,C	.0279	B,C:D
WTR1			
(1.50–1.94):(1.95–2.49)	B,C	.0092	B,C:D,E1,E2
(1.50–1.94):(1.95–2.49)	B,C	.0498	B,C:D,E1
(1.50–1.94):(1.95–2.49)	B,C	.0041	B,C:D
WI			
(0.95–1.04):(1.05–1.35)	D,E1,E2	.0018	B,C:D,E1,E2
(0.95–1.04):(1.05–1.35)	D,E1	.0024	B,C:D,E1
(0.95–1.04):(1.05–1.35)	D	.0163	B,C:D
LAC2			
(0.75–0.94):(0.95–1.39)	B,C	.0014	B,C:D,E1,E2
(0.75–0.94):(0.95–1.39)	B,C	.0045	B,C:D,E1
(0.75–0.94):(0.95–1.39)	B,C	.0139	B,C:D
HC			
(low):(high)	B,C	.0001	B,C:D

Note: For explanation of column headings see table 2.

TABLE 18
Significance of Differences in LP^3, LM^1, WM^1, NPC^1, NMC^1 and TC^2 in Allomyines from Intervals C, D, E1, E2 and F.

Variate and Intervals (mm)*	Values Higher at	Prob.	Stratigraphic Intervals
LP^3			
(1.00–1.39):(1.40–1.59)	C	.3333	C:D
(1.00–1.39):(1.40–1.59)	C	.1429	C:D,E1,E2,F
LM^1			
(1.90–2.19):(2.20–2.49)	C	.1000	C:D
WM^1			
(2.80–3.39):(3.40–3.69)	C	.0667	C:D
NPC^1			
(0–1):(2–5)	D	.0667	C:D
(0–1):(2–6)	D,E1,E2,F	.0065	C:D,E1,E2,F
NMC^1			
(1–2):(4–8)	D,E1,E2,F	.0083	C:D,E1,E2,F
TC^2			
(5–8):(9–23)	D,E1,E2,F	.0455	C:D,E1,E2,F

Note: For explanation of column headings see table 2.
*Except NPC^1, NMC^1 and TC^2, which are counts.

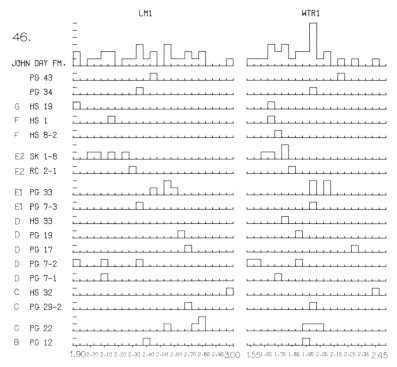

FIG. 46. Frequency distributions of length of M_1 (LM1) and width of trigonid on M_1 (WTR1) in allomyines from different locality-levels of John Day Formation. See also legend of fig. 22. Measurements in millimeters.

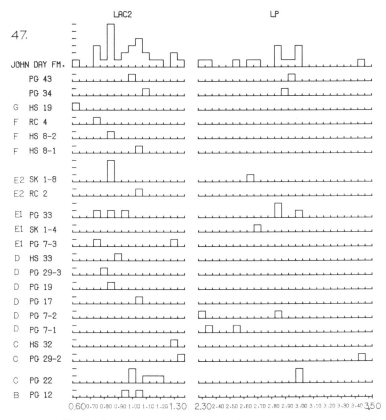

FIG. 47. Frequency distributions of length of anteroconid crest on M_2 (LAC2) and length of P_4 (LP) in allomyines from different locality-levels of John Day Formation. See also legend of fig. 22. Measurements in millimeters.

mesostylid is the most lingually prominent of the cusps on M_2 at each stratigraphic interval except G (pl. 11a) but has a distinctive prominence at B and C (pl. 11b).

The crest pattern of the upper cheek teeth resembles that in the prosciurines, consisting mainly of protoloph, metaloph, ectoloph and cingula. There are fewer accessory crests (NPC[1], NMC[1], TC[1] or TC[2]; fig. 52, table 18) than in specimens from higher intervals. In contrast, at interval D and higher (pls. 8a, 9a, 10f), small accessory crests extend from the primitive crests into the central valley.

Relationships. These differences indicate that the specimens from intervals B and C represent a population (group C) or closely related populations distinct from the groups occurring at higher stratigraphic positions. In view of the absence of accessory crests and the retention of strong resemblances of the protoloph and metaloph to morphologies in the prosciurine rodents, this group appears to represent a relatively primitive condition among the forms in the John Day Formation. However, as is observed below, the sample from interval D, which is more specialized in crest complexity, is more primitive than group C in crest height.

Group from Interval D

Beginning of a distinct type of lophodonty. In addition to the smaller size (see preceding discussions), the specimens of this interval differ from those of interval C in the basic pattern of lophodonty.

The occlusal crests of the specimens from interval D are much lower than in the specimens of group C. In most of the lower dentitions from interval D, the crests and cusps are worn almost flush with the centers of the valleys, yet some of the crests appear to be only lightly worn, with little if any dentine exposed along the crest axes. In similar stages of wear at interval C, crests show wider dentinal exposure yet elevations well above the bottoms of the valleys and fossettids. Two little worn specimens from interval D, UCMP 105025 (P_4-M_1) and UWBM 29297 (M_2) show that crests emanating from the apices of the major cusps descend down the flanks of the cusps without increasing in prominence (pl. 8b, viewed stereoscopically), whereas those of Group C increase in prominence above the tooth surface with greater distance from the origin (pl. 11b). As a consequence, at the centers of the crowns, the crests in the specimens from interval D are much less elevated than those of group C. The effect of this lesser elevation in the specimens from interval D is the retention of a more pronounced occlusal relief, like that characteristic of the prosciurine dentition, whereas the relief in group C is flatter, even in unworn stages, like the occlusal surfaces in truly hypsodont rodents.

In addition to their lower height, the crests at interval D have an orientation distinct from those of group C. The internal crests of the anterior part of the molar at interval D tend to converge toward the center of the trigonid basin (pl. 8b) rather than trend posterolinguad as in group C (pl. 11c). In the specimens of interval D, the smaller size of the anteroconid crest (LAC2; fig. 47, table 17), which in group C occupies much of the trigonid basin, permits convergence of other crests into the open space. Accessory crests in the upper cheek teeth are more numerous than in group C (TC[1], TC[2], NPC[1], NMC[1], PAPR[1]; figs. 52-53, tables 18-19. These incipient crests, most of which emanate from the protoloph and metaloph, are directed toward the centers of the valleys. The protoloph and metaloph are better defined as transverse lophs and the

TABLE 19
Significance of Differences in TC[1], NMC[1], PAPR[1], TC[2] and MLC[1]
in Allomyines from Intervals C and D.

Variate and Intervals (mm)*	Values Higher at	Prob.	Stratigraphic Intervals or locality
TC[1]			
(8–9):(10–18)	D	.1667	D:C
NMC[1]			
(1–2):(4–8)	D	.1667	D:C
PAPR[1]			
(absent):(present)	D	.1000	D:C
TC[2]			
(5–8):(9–15)	D	.1429	D:C
(5–8):(9–15)	PG7(1,2)	.1000	PG7(1,2):C
MLC[1]			
(0.00–0.09):(0.10–0.34)	D	.0667	D:C

Note: For explanation of column headings see table 2.
*Except TC[1], NMC[1] and TC[2], which are counts.

TABLE 20
Significance of Differences in LM1, WTR1, LM[1], WM[1], HC,
TCDP[4], NMC[1] and MLC[1] in Allomyines from Intervals C, D, and E1.

Variate and Intervals (mm)*	Values Higher at	Prob.	Stratigraphic Intervals
LM1			
(1.90–2.39):(2.40–3.09)	E1	.1753	E1:D
WTR1			
(1.50–1.94):(1.95–2.49)	E1	.0075	E1:D
LM[1]			
(1.90–2.29):(2.30–2.69)	E1	.0286	E1:D
WM[1]			
(2.80–3.29):(3.30–3.69)	E1	,0143	E1:D
HC			
(low):(high)	E1	.0056	E1:D
TCDP[4]			
(0–14):(15–16)	E1	.3333	E1:D
NMC[1]			
(1–2):(3–6)	E1	.0278	E1:C
MLC[1]			
(0.00–0.09):(0.10–0.34)	E1	.0476	E1:C

Note: For explanation of column headings see table 2.
*Except TCDP[4] and NMC[1], which are counts.

transverse valleys are more distinct than in group C. The protoloph and metaloph exhibit less anterior and posterior excursion across the molars, that is, they are straighter, as shown by the anteroposterior separation of the labial bend of the metaloph from the center of the central valley on M^1 (MLC^1) in table 19.

The major cusps of the protoloph and metaloph are less distinct or isolated on the lophs than in group C, so that the fossettes separating the cusps of each loph are transversely narrower and less conspicuous than in group C. The protoconule is barely distinguishable from the rest of the protoloph, especially on M^2. The fossettes of the metaloph in the molars are bounded by the flat surfaces of the metaconules and protocone, giving these depressions the appearance of narrow, clearly defined anteroposteriorly aligned depressions. The flat labial and lingual surfaces of the conules give these cusps a more pronounced quadrangular shape than is typical of group C.

The anterior and posterior surfaces of wear on protoloph and metaloph are more steeply inclined, both toward the central valley and toward the anterior and posterior margins of the teeth, than in group C. This results in a deeper definition of the central valley and the valleys between the teeth.

The differences in crest development in groups C and D seem to represent different space-filling strategies. In group C, the greater separation of the cusps had been accompanied primitively by longer crest connections forming the protoloph and metaloph. These connections already had some lateral excursion (the primitive zigzag aplodontid pattern, seen in both European and North American taxa). The increase in lophodonty leading to group C simply emphasized the degree of excursion of these parts of the protoloph and metaloph in order to utilize the space in the transverse valleys. In group D, however, the closer union of the cusps of the protoloph and metaloph left the connecting crests of the lophs very small. Consequently, under selection for lophodonty, the more completely open valleys could be more rapidly invaded by new crests at new points along the margins of the valleys. Group D represents an early stage in this type of lophodonty and seems to be close to the morphologic starting point for all the later allomyine groups of the John Day Formation.

Divergent specimens from interval D. Most specimens from interval D were found at Picture Gorge 7 (V–6506, UWA 5183). Some specimens from other localities of this interval differ somewhat from the forms of Picture Gorge 7.

A single mandible from locality Picture Gorge 17 (UWA 5171) differs from those of levels 1 and 2 at Picture Gorge 7 (V–6506, UWA 5183). The specimen from Picture Gorge 17 (UWBM 43375) is the closest of all the specimens to the group from interval C. However it differs from all those unequivocally from beneath the Deep Creek tuff in the length and attitude of the internal protoconid crest, which is short and directed toward the mesostylid rather than the entoconid. This specimen occurred only a very short distance above the Deep Creek tuff. An upper dentition, UWBM 43278, from Picture Gorge 17 is larger than and lacks most of the accessory crests characteristic of the specimens from Picture Gorge 7 and resembles group C in these features. These specimens are important because they substantiate the apparent continuation of group C to the lowest of the deposits above the Deep Creek tuff.

A mandible from level 2 of Picture Gorge 29 (UWA 9596) occurred on the upper surface of a bed that probably represents the Deep Creek tuff (based on occurrences of *Meniscomys,* p. 86). In this specimen (pl. 11c), the internal protoconid crest is shorter than in group C, but as in the latter, is directed toward the entoconid.

A mandible from Picture Gorge 19 (A-9592) is larger and has higher crests than are typical of group D but lacks the elongation and heaviness of the crests of group C. This specimen, UWBM 43105, is distinct from the specimens of the overlying intervals E1 and above in the greater degree of lingual convexity of M_1 near the mesostylid.

Two maxillary dentitions from Picture Gorge 19 are also larger than those typical of interval D, although, like others of this interval, both exhibit accessory crests extending into the central valley. UWBM 43112 bears lower and more slender major crests than those of group C, and the crests of the protoloph and metaloph lack the anteroposterior excursion characteristic of group C. This specimen seems most closely related to the other specimens of interval D but is somewhat advanced. The crests of UWBM 43107 are more prominent than those typical of interval D. The sample from Picture Gorge 19 therefore seems to be most closely related to group D at Picture Gorge 7 but is more advanced in size and crest height.

In summary, most of the specimens collected from interval D represent a population (group D) morphologically distinct from group C and characterized by smaller size and the primitively low and distinct structure of the lophs and accessory crests. This group contains the collection from levels 1 and 2 of Picture Gorge 7 and is probably represented in a somewhat advanced state by the specimens from Picture Gorge 19. The different crest arrangement and lower crest height show that group D does not represent a continuation of the underlying group C but is a distinct lineage.

Group from Interval E1

Size increase. The cheek teeth in the specimens from the stratigraphic interval above the *Meniscomys hippodus* teilzone and beneath the *Pleurolicus sulcifrons* teilzone (fig. 45) are larger than those of interval D. For example, these specimens differ significantly from those of interval D in the length of M_1 (LM1), the width of the trigonid on M_1 (WTR1), the length of M^1 (LM^1), and the width of M^1 (WM^1), as shown in table 20 and figures 46 and 51. However, although overlap exists, the size of the specimens at interval E1 does not attain values as great as in group C.

Crest prominence. The prominence of the crests at interval E1, as indicated by general height (HC in table 20 and fig. 49) is significantly greater than at interval D. Even in a heavily worn stage, the lower teeth at Picture Gorge 33 of interval E1 show distinct fossettids with vertical walls (pl. 9c), whereas the fossettids of group D are vaguely defined due to less frequent closure and gently sloping walls (pl. 8b). Though difficult to quantify, the general height of the crests is not as great as at interval C.

Crest complexity and orientation. The number of accessory crests in two specimens of DP^4 found at interval E1 (pl. 9b) is much greater (15–16) than in the single specimen (5) found at interval D (TCDP^4 in fig. 53), and the single specimen (4) found at interval C (pl. 12b).

No significant difference in the number of crests in the permanent teeth between intervals E1 and D was found. However, the number of accessory crests in the permanent cheek teeth at interval E1 is significantly greater than at interval C, as is shown, for ex-

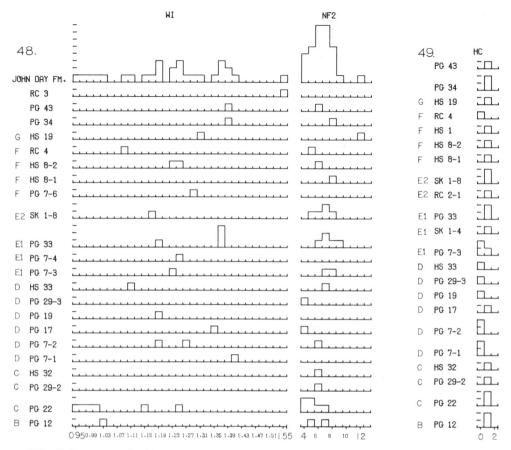

FIG. 48. Frequency distributions of width of lower incisor (WI) and number of fossettids of M_2 (NF2) in allomyines from different locality-levels of John Day Formation. See also legend of fig. 22. Measurements of WI in millimeters.

FIG. 49. Frequency distributions of relative crest height (HC) in cheek teeth of allomyines from different locality-levels in John Day Formation. See also legend of fig. 22. Measurements in millimeters.

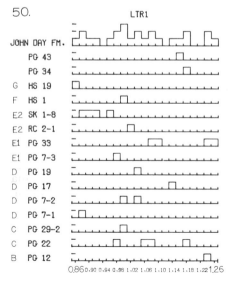

FIG. 50. Frequency distributions of length of trigonid on M_1 (LTR1) in allomyines from different locality-levels of John Day Formation. See also legend of fig. 22. Measurements in millimeters.

ample, by the count of crests on the anterior side of the metacone and metaloph of M^1 (NMC^1 in table 20 and fig. 52).

The metastylid crest is aligned more strongly anteroposteriorly and the lingual outline is less convex than in group C. The internal crest of the protoconid on M$_2$ is directed toward the mesostylid (pl. 9c), not the entoconid (pl. 11c) as in group C, and intersects the anterostylid crest somewhere in front of the posterior end of that crest. The fossettids of the trigonid tend to be rather oval, not anteroposteriorly elongate as in group C, and this feature permits distinguishing even heavily worn dentitions of these groups. There is considerable variability in the number of accessory crests in the lower dentition at Picture Gorge 33, even between the left and right rami of a single individual, but every individual shows greater complexity than exists in group C.

The crests joining the cusps of the metaloph and protoloph do not extend as far into the central valley (pl. 9a,b) as in the specimens of group C (pl. 12a,b): for example, the separation of the labial bend of the metaloph from the middle of the central valley (MLC^1 in table 20) is significantly greater at interval E1.

These relationships indicate that interval E1 contains specimens that are distinct from those of either of the lower stratigraphic intervals. However, more than one morphological group seems to be present in this sample.

Transitional samples from Picture Gorge 7, level 3. The three upper dentitions from level 3 of Picture Gorge 7 (V–6506, UWA 5183) are probably more closely related to specimens of group D than to those from other localities of interval E1. Two of these specimens consist of M^3, which is not otherwise known for group D. However, these specimens are of the small size expected in group D, exhibit low crests and accessory crests and have an almost imperceptible protoconule on a uniform protoloph, as in group D. Although M^3 is not known from other localities of interval E1, the protoconule and metaconule are well defined in each of the four specimens with M^1 or M^2 at the other localities of this interval. The third specimen from level 3 at Picture Gorge 7 is an M^2, which is a little larger than the specimens of group D but resembles the latter in other features, including poor definition of the protoconule. These specimens seem therefore to represent group D.

Two lower dentitions from level 3, UCMP 105026 and 105028, resemble specimens typical of interval E1 in larger size, but bear crests not much more prominent than those of group D. An M$_2$ from level 3, UCMP 97074, is still larger than the teeth of UCMP 105026 and 105028 and bears crests as high as in any specimen from interval E1. It appears that the lower part of interval E1, as represented by level 3 of Picture Gorge 7, contains a population related to, though perhaps more advanced than, group D and another population related to specimens from other localities of interval E1. Specimen UWBM 43112 from Picture Gorge 19 (UWA 9592) may represent the upper dentition of this advanced group D.

Relationships. The small accessory crests emerging from the main transverse lophs of the upper molars in group D represent an incipient lophodonty, independent of the simpler more heavily crested lophodonty present in group C (see p. 95).

In group E1, the crests of the primitive transverse lophs, though more excursive than in group D, do not extend as far into the central valley as in group C. Instead, low, short accessory crests not present in group C extend from the protoloph and metaloph into the central valley. The anterior cingulum flares anteriorly more conspicuously than in group C. As a consequence of this and the smaller protoconule, the anterior transverse valley is wider and less obstructed than in group C.

These conditions generally derive from those of group D, for they represent a partial filling of primitively open transverse valleys by the slight enlargement of structures present in group D. In these changes, group E1 has converged on the structural adaptations of group C, though not as advanced in size, crest height, crest length, horizontal wear, or filling of the transverse valleys, yet has retained the accessory crests of group D. The presence of several specimens somewhat intermediate between groups D and E1 in the lower part of interval E1 favors this interpretation. Group E1 seems to represent a continuation of the trend toward increased lophodonty of a special type that began in group D, a trend functionally convergent on that already attained to a high degree by group C. In both groups D and E1 the lophodonty involves filling space with small crests as opposed G to lengthening the existing primitive crests as in the inferred transition to group C.

However, group E1 may not be derived from group D as known. In group E1 the primitive crests of the protoloph and metaloph, though more restricted than in group C, are better developed than in group D, and the upper molars are relatively longer anteroposteriorly. Although the structure of the protoloph and metaloph could be derived from group D, the condition in group E1 looks more like the inferred primitive condition, which occurs in group C as well as in meniscomyines and prosciurines, whereas the condition in group D seems modified from the primitive condition. The increase in anteroposterior length could also be derived from group D, but this change seems functionally unrelated to the lophodont trend because the increase in length leaves increased unfilled space, especially in the anterior transverse valley. Group E1 may therefore be derived from another locally unrepresented population or subpopulation in which the upper molars were longer and the loph condition somewhat more primitive than in group D.

Group from Intervals E2 and F

Small size. The cheek teeth in the specimens from intervals E2 and F are smaller than in the collections from interval E1. The length of M_1 (LM1), the width of the trigonid on M_1 (WTR1), and the length of the trigonid on M_1 (LTR1) are significantly smaller at intervals E2 and F (table 21 and figs. 46, 50). The length of M^1 (LM^1) and the width of M^1 (WM^1) are significantly less at intervals E2 and F (table 21 and fig. 51) than at interval E1. The size in these specimens extensively overlaps the range in size at interval D. The small size of these specimens also serves to differentiate them from group C as, for example, in length of M_1 (LM1 in table 21 and fig. 46).

Crest height and complexity. The height of the crests at intervals E2 and F is as great as that at interval E1 and significantly greater than that at interval D (HC in table 21 and fig. 49).

Although the height may be similar, the configuration of the crests in the lower cheek teeth at intervals E2 and F is distinct from that at interval E1. The smaller of the accessory crests tend to have disappeared or attained a thickness and height similar to that of the major crests. Together the crests tend to fill available space more uniformly than those at interval E1. For example, the internal metaconid crest on M_2 or M_3 is usually present (pl. 10d) at intervals E2 and F but more often absent or very weak (pl.

TABLE 21

Significance of Differences in LM1, WTR1, LTR1, LM[1], WM[1], WM[2], HC, IMC2, IPC2 and ACM[P] in Allomyines from Intervals B, C, D, E1, E2 and F.

Variate and Intervals (mm)	Values Higher at	Prob.	Stratigraphic Intervals
LM1			
(2.00–2.29):(2.30–2.69)	E1	.0022	E2,F:E1
(2.00–2.39):(2.40–3.09)	B,C	.0006	E2,F:B,C
WTR1			
(1.50–1.94):(1.95–2.49)	E1	.0013	E2,F:E1
LTR1			
(0.90–0.99):(1.00–1.29)	E1	.0454	E2,F:E1
LM[1]			
(1.90–2.49):(2.50–2.79)	E1	.0079	E2,F:E1
WM[1]			
(2.80–3.29):(3.30–3.59)	E1	.0079	E2,F:E1
WM[2]			
(2.65–3.14):(3.15–3.44)	E1	.1667	E2,F:E1
HC			
(low):(high)	E2,F	.0004	E2,F:D
IMC2			
(absent, weak):(strong)	E2,F	.1573	E2,F:E1
IPC2			
(turns posterolabiad): (turns posterolinguad)	E2,F (labiad)	.0862	E2,F:E1
ACM[P]			
(0.00–0.09):(0.10–0.29)	F	.0476	F:E2,E1,D,C

Note: For explanation of column headings see table 2.

9c) at interval E1 (IMC2 in table 21). The internal protoconid crest of M_2 most often curves posteriad, joins the anterolingual crest of the mesoconid, and forms a short, oval fossettid at intervals E2 and F. At interval E1 this crest is most often straight or turns anteriad so that, if a fossettid is formed between it and the mesoconid, the structure is more elongate than at E2 and F (see character IPC2 in table 21).

As in the lower dentition, there is a tendency in the upper molars for the crests to fill space more uniformly (pl. 10d). In UCMP 79895 and 105043, this involves slight displacement, enlargement, or reduction of accessory crests, producing a random space filling pattern. In UCMP 105039, the anterior crests of the protoconule and paracone tend to extend toward and reach the anterior cingulum and occupy the anterior transverse valley (pl. 10f). In UCMP 105033, accessory crests are less numerous but the anteroposterior shortness of the molars has reduced the intercrest space, especially in the anterior transverse valley.

There is a tendency in the upper cheek teeth at intervals E2 and F for the anterior half of the ectoloph to be rather deeply invaginated, especially adjacent to the basal cingulum and the mesostyle. P^4 at intervals E2 and F differs significantly in this respect from specimens of lower stratigraphic positions (ACM^P in table 21 and fig. 54). In the upper molars, the invagination is somewhat less pronounced.

Relationships. The specimens of intervals E2 and F represent a group (F) with consistent differences from the specimens of underlying positions. However, these differences are of degree and represent extensions of features present in one or the other of the underlying groups. A reduction in size as well as some modification of the crest development would have accompanied derivation from group E1. Little or no increase in size but a greater amount of crest modification would have been involved in a transition from group D. A tendency for the upper molars to be shorter relative to transverse width is shared with group D but not group E1. The evidence therefore seems to favor an origin for group F from a population like that of group D, even though group D had locally disappeared during interval E1.

Specimen from Interval G

Shorter cheek teeth. A single mandibular fragment of an allomyine was found in the teilzone of *Entoptychus individens,* the terminal species of *Entoptychus,* at the top of the *Entoptychus-Gregorymys* Concurrent-range Zone (fig. 45). This specimen differs from the others of the formation, taken collectively, in the anteroposterior shortness of the cheek teeth. The M_1 (LM1 in table 22 and fig. 46) and the trigonid on M_1 (LTR1 in table 22 and fig. 50) at Haystack 19 (V–6587) are significantly shorter than in the specimens from the lower stratigraphic positions taken together, although in group F the cheek teeth may be equally narrow transversely. The M_3 in particular is much shorter than in other specimens.

Shorter, anastomosing crests. The anteroconid crest on M_2 (LAC2 in table 22 and fig. 47) is shorter at Haystack 19 than in the specimens from lower stratigraphic positions (pl. 11a). The number of fossettids on M_2 (NF2 in table 22 and fig. 48) is greater than in the other specimens, owing to increased union of crests.

The internal protoconid crest of M_2, which tends to curve posterolabiad in group F to enclose an oval rather than an elongate fossettid, here forms a tiny, essentially cir-

TABLE 22

Significance of Differences in LM1, LTR1, LAC2 and NF2 in an Allomyine from Interval G and Allomyines from Intervals B, C, D, E1, E2 and F.

Variate and Intervals (mm)*	Values Higher at	Prob.	Stratigraphic Intervals
LM1			
(1.80–1.89):(2.00–2.29)	E2,F	.1429	G:E2,F
(1.90–1.99):(2.00–3.09)	B,C,D,E1,E2,F	.0400	G:B,C,D,E1,E2,F
LTR1			
(0.80–0.89):(0.90–1.09)	E2,F	.1429	G:E2,F
(0.80–0.89):(0.90–1.29)	B,C,D,E1,E2,F	.0454	G:B,C,D,E1,E2,F
LAC2			
(0.60–0.69):(0.70–1.09)	E2,F	.1250	G:E2,F
(0.60–0.69):(0.70–1.39)	B,C,D,E1,E2,F	.0385	G:B,C,D,E1,E2,F
NF2			
(5–11):(12)	G	.1111	G:E2,F
(4–11):(12)	G	.0400	G:B,C,D,E1,E2,F

Note: For explanation of column headings see table 2.
*Except NF2, which is a count.

cular lake (pl. 11a). A crest from the entoconid reaches the internal protoconid crest. The anterointernal crest of the entoconid is more strongly joined to the mesostylid than in other specimens, and the connecting crest is elevated so that it early attains the plane of wear. The crests in general seem to be more uniform in height, producing a flatter occlusal surface than in the other specimens of a similar stage of wear (pl. 11a). In specimens with as flat occlusal surfaces on M_{1-2}, many of the crests have been completely worn away.

The lingual margins of M_{1-2}, which are most convex in group C and less convex in groups D, E1, and F, are straighter in this specimen than in any of the others. The lingual surface of the metastylid crest, between metaconid and mesostylid, is more sharply invaginated than in other specimens. A pronounced lip or cingulum, which forms the anteroventral boundary of this depressed area, is absent at lower stratigraphic positions. The mesostylid and the entoconid are more nearly oval in occlusal outline than in other groups.

Relationships. This specimen represents a group in which smallness, especially shortness of cheek teeth, shorter crests and union of crests to form small, oval fossettids, has reached a rather advanced stage. The closure of fossettids represents a continuation of the space-filling trend that characterizes the succession from group D through group F. The small size of this specimen is also characteristic of the underlying group F. The shortness of the lower cheek teeth is the only characteristic that may not be consistent with an origin from group F. Nevertheless, the specimen from interval G has a greater overall similarity to group F than to any of the stratigraphically more distant groups of allomyines.

Specimens from Unknown Stratigraphic Positions

Specimen from Rudio Creek 3. An allomyine from Rudio Creek 3 (V–66106) most closely resembles the specimens from interval E1, especially those from Haystack 6 (V–6590), in size and intermediate crest height. The lower cheek teeth in this specimen, UCMP 105041, are worn to the same late stage as those of Haystack 6 and the slopes and diameters of the fossettid bases are very similar. The upper cheek teeth resemble those of group E1 in the limited development of accessory crests. The crests in these specimens were higher than those of interval D but not as high as those of interval C. The larger size of the cheek teeth also distinguishes this specimen from those of intervals D, E2 and F. It seems most likely that this deposit represents interval E1. This interpretation is consistent with the absence of any indication of *Entoptychus,* the teilzone of which overlies intervals E1 and E2, the absence of *Pleurolicus,* which overlies E1, and the absence of *Meniscomys,* which only occurs as high as the lower part of interval E1. The biostratigraphic relationship of this site is important because it is the only locality at which the unusual meniscomyine *Rudiomys mcgrewi* has been found.

Specimens from Picture Gorge 34. Two specimens from Picture Gorge 34 (UWA 31473) also represent group E1. Crests are of intermediate height, and the size is greater than that in specimens from either interval D or intervals E2, F or G.

Specimen from Picture Gorge 43. The single specimen, UWBM 47310, from Picture Gorge 43 (UWA 9966) likewise represents group E1. It occurred above the Picture Gorge ignimbrite in a buff colored deposit that resembles the sediments at Picture Gorge 33 (interval E1 by biostratigraphic evidence) and Picture Gorge 34, all of which are located in an area with a radius of about 1.6 km.

Specimen from Picture Gorge 6. An upper molar, UWBM 48160, was recovered from 4 m above the base of the exposure at Picture Gorge 6 (UWA 9581). This specimen most closely resembles the molars of group E1.

Group Equivalence in Systematic Section

These groups of allomyines, differing in morphology and stratigraphic or geographic separation, are identified by the following taxonomic names in the section on systematics.

Group	Taxon	Intervals
G	*Allomys tessellatus,* n. sp.	G
F	*Allomys reticulatus,* n. sp.	F, E2
E1	*Allomys nitens* Marsh	E1
D	*Allomys simplicidens,* n. sp.	E1, D, D-*
C	*Alwoodia magna,* n. gen., n. sp.	D-,* C+,* C, B

*D- = a few feet above Deep Creek tuff
C+ = upper surface of Deep Creek tuff

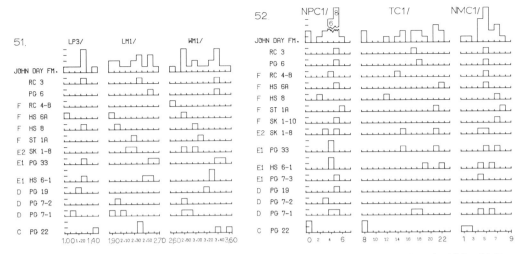

FIG. 51. Frequency distributions of length of P³ (LP³), length of M¹ (LM¹) and width of M¹ (WM¹) in allomyines from different locality-levels of John Day Formation. See also legend of fig. 22. Measurements in millimeters.

FIG. 52. Frequency distributions of number of accessory crests on posterior surface of protoloph area of M¹ (NPC¹), total number of accessory crests on M¹ (TC¹), and number of accessory crests on anterior surface of metaloph area of M¹ (NMC¹) in allomyines from different locality-levels of John Day Formation. See also legend of fig. 22.

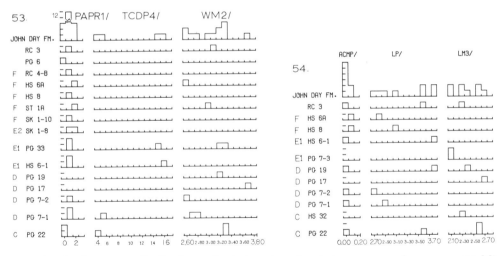

FIG. 53. Frequency distributions of relative size of anterior process of paracone on M¹ (PAPR¹), total number of accessory crests on DP⁴ (TCDP⁴) and width of M² (WM²) in allomyines from different locality-levels of John Day Formation. See also legend of fig. 22. Measurements of WM² in millimeters.

FIG. 54. Frequency distributions of depth of anterior concavity of mesostyle on P⁴ (ACMᴾ), length of P⁴ (LP⁴), and length of M³ (LM³) in allomyines from different locality-levels of John Day Formation. See also legend of fig. 22. Measurements in millimeters.

Summary of Intrabasinal Relationships of Meniscomyines and Allomyines

The earliest meniscomyine or allomyine to appear in the John Day Formation is *Meniscomys,* below but near the base of the Picture Gorge ignimbrite (fig. 21).

Allomyines first appear in the beds immediately above the Picture Gorge ignimbrite in association with occurrences of *Meniscomys* (fig. 45). The deposition of the Picture Gorge ignimbrite probably occurred in a few months or less; thus, relatively little geological time elapsed between the apparent appearances of the allomyines and meniscomyines. It is quite possible that allomyines and meniscomyines appeared almost simultaneously, because the only site where a meniscomyine predates the allomyines has produced only a single specimen of *Meniscomys* and gives no proof that allomyines were not present. From above the the Picture Gorge ignimbrite upward, throughout the teilzones of *Meniscomys* and *Rudiomys,* the meniscomyines were sympatric with the allomyines (fig. 45). Differences in cranial and postcranial structures, which may provide indications of adaptive differences between the meniscomyines and allomyines, will be discussed in a separate study.

Meniscomyines. The earliest recognized meniscomyine species, *Meniscomys uhtoffi,* ranges from the strata near the base of the Picture Gorge ignimbrite upward to the base of the Deep Creek tuff (fig. 21). Although the samples of *Meniscomys* from this interval vary slightly in crown height, there is no evidence of increasing hypsodonty with increasing height in the section. Some very brachyodont specimens occur quite close to the Deep Creek tuff. Some of the specimens from this interval exhibit a distinct mesostylid cusp on P_4 (figs. 2, 4), whereas in several specimens from apparently the same stratigraphic position, there is only a uniform metastylid crest in that area of the tooth (pl. 2h,i). A stylid appears on the anterior surface of the P_4 hypoconid (HSHP in fig. 3) in this species at higher stratigraphic positions, but the difference is not significant. Neither of these structures contributes to hypsodonty.

Meniscomys hippodus appears a short distance above the Deep Creek tuff (fig. 21), and differs from *M. uhtoffi* in a number of characters, including an increase in hypsodonty of at least the lower part of the crown. The heights of enamel inflections above the base of the crown are greater in *M. hippodus* (fig. 5). The sides of the labial inflection of P_4 are more vertical, forming a narrower angle where they converge at the base of the crown (pl. 2e). The increase in occlusal length of the mesostylid crest on P_4, the anterior spur on the hypoconulid of P_4, the length of M_1 and the length of the trigonid on M_2 contribute to increased lophodonty (fig. 5). The anteroposterior length of the cross section of the lower incisor of *M. hippodus* is greater than that of *M. uhtoffi.*

M. hippodus is the most abundantly represented species of *Meniscomys*. The larger count indicates larger numbers of individuals in the original populations rather than a bias in the preservation or sampling processes, judging from the wide geographic distribution of the localities where this is the most common species. This distribution includes Haystack Valley, Picture Gorge quadrangle, and the Camp Creek region 100 km south of Picture Gorge (fig. 1).

At the locality where the greatest vertical distribution of *M. hippodus* occurs, Picture Gorge 7 (V–6506, UWA 5183), the species shows greater morphologic variation at the higher end of its range (e. g., LIAP in fig. 29). In a number of characters specimens of level 2 exhibit states not present at the lower level. The morphologic ranges of some characters at level 2 more strongly overlap the ranges in *M. uhtoffi* than do those at level 1. The sample from level 2 therefore seems to contain a heterogeneity not easily explained as simply derived from the observed variation of the earlier members of the species. The samples from levels 1 and 2 at Picture Gorge 7 also differ in certain characters from approximately correlative samples at other localities, especially Picture Gorge 17 and Schrock's 1.

The abundance, widespread distribution, heterogeneity and local reversals of character trends imply that *M. hippodus* was a widespread species of great variability and at single geographic positions was represented by a succession of temporally short-lived local populations, at least some of which were replaced by local populations that had existed elsewhere in sufficient isolation to acquire distinct characteristics.

Meniscomys editus occurs stratigraphically above *M. hippodus* at Picture Gorge 7 (fig. 21), is represented by fewer specimens than either of the earlier species, and is unknown at other localities. The poor record of this species is probably an indication of a smaller population size, unless factors affecting preservation deteriorated on a regional basis. A number of other fossiliferous exposures, from Haystack Valley through Picture Gorge quadrangle to the Camp Creek region, bracket this interval, that is, contain both the underlying teilzone of *M. hippodus* and the overlying teilzone of *Pleurolicus,* yet have not produced *M. editus*. There is no evidence of an unconformity through this interval and at one site, Schrock's 1, the section overlying the *M. hippodus* teilzone (Rensberger, 1973b:fig. 4) is thick and has yielded other rodents but not meniscomyines.

M. editus is advanced in many characters over its predecessors. These differences, like those separating *M. uhtoffi* and *M. hippodus,* involve increases in crown height and lengthening of crests. The most prominent change in *M. editus* was the development of higher, more vertically bordered dentinal tracts on the anterior and posterior sides of the cheek teeth, although this feature is rarely seen because it occurs between adjacent teeth. But other indications of advanced hypsodonty exist, including lengthened crowns (fig. 9), height of the premolar mesostylid (fig. 8) and stylid on the hypoconid (fig. 9), increased thickness of incisor enamel, and increased parallelism of the sides of the crowns, as exemplified by a decrease in the angle between the walls of the labial inflection on P_4 (fig. 12). The cheek teeth of *M. editus* are larger than those of *M. hippodus* (fig. 11). However, the changes in hypsodonty represent larger percentages of change than do other measures of size, such as the anteroposterior diameter of the tooth.

The greater lophodonty in *M. editus* is indicated by the transverse flattening of certain lophs. As in changes involving lophodonty in other herbivorous mammals, the trend involves accentuation of lophs facing the direction of occlusal motion, which in aplodontids is nearly transverse. This can be seen in the flattening of the primitively oval paracone and in the disproportionate anteroposterior lengthening of the protocone or lingual moiety of the upper cheek teeth at the expense of the transverse dimension (fig. 7).

The succession of closely related species, *Meniscomys uhtoffi*, *M. hippodus* and *M. editus*, apparently terminated with *M. editus*. Changes in crown height and lophodonty through this succession resemble dental changes that later characterized the history of *Entoptychus* as it became increasingly fossorial (Rensberger, 1971).

Rudiomys mcgrewi occurs stratigraphically above the final appearance of *Meniscomys*, a position established by an associated occurrence of *Allomys nitens*, which consistently underlies the *Pleurolicus* teilzone and is not found in association with *Meniscomys* (fig. 45). The *Rudiomys* site was probably topographically higher than other occurrences of meniscomyines during John Day time. *Rudiomys* differs from *Meniscomys* in its more brachyodont but occlusally larger cheek teeth, radial rather than vesicular partitioning of the bulla, and arched rather than straight mesostylid crest (pl. 5a,c,e).

No remains of *Sewelleladon predontia* other than the type are known. The type locality is probably within the Haystack Valley Member, in the *Mylagaulodon* Concurrent-range Zone or deposits just beneath that unit. Although *Sewelleladon predontia* is much advanced in lophodonty and hypsodonty (pl. 6a,b), it may bear a special relationship to *Rudiomys mcgrewi*. These species both postdate *Meniscomys*, and share morphologic similarities reviewed in the following section. *R. mcgrewi* and *S. predontia* may have been competitive with *Meniscomys* and *Entoptychus*, respectively, for they appear only after termination of long sequences of those taxa (fig. 45). *Rudiomys mcgrewi*, which occurs between those sequences, has never been found in the beds above, where several thousand specimens of rodents, dominantly *Entoptychus*, have been recovered. The magnitude of morphologic change accrued by *Entoptychus* during the period intervening between the occurrences of *R. mcgrewi* and *S. predontia* is comparable in magnitude to the morphologic differences separating these aplodontid species.

Allomyines. The allomyine occurring lowest in the section, *Alwoodia magna*, ranges from a short distance above the Picture Gorge ignimbrite to a few feet above the Deep Creek tuff (fig. 45), a range which closely matches that of the earliest species of *Meniscomys*, *M. uhtoffi* (fig. 21). *Alwoodia magna* differs from other allomyine species in the large size of the cheek teeth, yet narrower incisor; heavier, higher and more elongate crests in the cheek teeth; and absence of small, accessory crests (pls. 11b-d, 12). There is also a greater degree of parallelism among crests of the lower molars than in other species. In terms of crest height and length, *Alwoodia magna* is more advanced than any other allomyine species except perhaps the terminal one, whereas the entoptychines and meniscomyines first appeared with clearly more primitive morphologies than later related forms. However, *Alwoodia magna* is more primitive than later allomyines in certain characters. The lingual border of the lower molars is strongly convex, a characteristic of early species of *Allomys* and of the European species, *Parallomys ernii,* in which accessory crests are poorly developed.

Allomys simplicidens first appears a short distance above the Deep Creek tuff, at approximately the position of highest occurrence of *Alwoodia magna,* but without evidence of overlap at any single locality (fig. 45). This species represents the appearance of a new lineage, because the crests of the molars are extremely low and short and the occlusal size of the cheek teeth is much smaller (pl. 8) than that of *Alwoodia magna.* Furthermore, the crests of the trigonid lack the tendency for parallelism exhibited by *Alwoodia magna,* and tend instead to converge toward the center of the basin (pl. 8b). The lower molars are truely basined, for the crests are low and tend to descend down the sides of the bordering cusps.

In the upper molars (fig. 14a) very short, low crests emerge from the protoloph and metaloph and initiate new structures and a distinct type of lophodonty not present in the prosciurines or *Alwoodia.* The anterior and posterior transverse valleys in the upper molars are not obstructed by either low accessory crests or the very short arms of the zigzagging protoloph and metaloph. These valleys retain a distinct differential relief into late stages of wear. In other words, the upper molars retain much of the functional occlusal anatomy characteristic of aplodontids of prosciurine grade, the main differences being the stronger development of the ectoloph and the presence of a double metaconule. The labial metaconule (fig. 13) probably goes back to an almost completely prosciurine grade of development, when cusps alone formed the relief of the upper cheek teeth.

Allomys cavatus, known only by the type from an unknown locality, is even more primitive than *Allomys simplicidens.* The cheek teeth are still smaller than those in *Allomys simplicidens,* the secondary crests are smaller, being best described as crenulations, and the lingual margin of the lcwer molars is more convex (pl. 7a,b). These features suggest that *Allomys cavatus* represents a form close to the ancestry of *Allomys simplicidens.* However, in the Picture Gorge quadrangle where *Allomys simplicidens* is found, there is no barren stratigraphic interval nor any evidence of an unconformity between the occurrences of *Allomys simplicidens* and its predecessor, *Alwoodia magna,* that would suggest that *Allomys cavatus* may be locally present beneath the occurrence of *Allomys simplicidens.* Several features, including relatively narrower cheek teeth and greater confinement of wear to the apices than to the sides of the conules of the upper cheek teeth, may indicate that *Allomys cavatus* was not the direct ancestor of *Allomys simplicidens.* Species with morphologies close to that of *Allomys cavatus* almost certainly existed somewhere at the time of occurrence of *Alwoodia magna,* and the type of *Allomys cavatus* may represent one such population.

Allomys nitens occurs in deposits stratigraphically above those of *Allomys simplicidens* and ranges up to but not into the *Pleurolicus* teilzone (fig. 45). Cheek teeth of *A. nitens* are larger than those of other species of *Allomys,* but not as large as those of *Alwoodia magna* (pl. 9). Secondary crests are higher and often longer than in *Allomys simplicidens* but do not extend to fill the central, anterior or posterior transverse valleys (pl. 9a). However, the expansion of crests is sufficient to reduce the distinctiveness of the protoloph and metaloph, so that the dominant functional structures in the interior of the upper cheek teeth are the small crests rather than the edges of the protoconule and metaconule. The total number of accessory crests in the permanent cheek teeth is not greater than in *Allomys simplicidens,* but the number in DP^4 is increased (pl. 9b). The anteroposterior length of the molars is greater in relation to width than in *Allomys simplicidens,* a proportion more like that in *Allomys cavatus.*

Allomys reticulatus appears in the teilzone of *Pleurolicus* and is found well into the *Entoptychus* teilzone (fig. 45), at least as high as *E. cavifrons.* The cheek teeth are smaller than in *Allomys nitens* (pl. 10d,e,f), overlapping in size those of *Allomys simplicidens.* Crests are not noticeably higher than in *Allomys nitens* but are less irregular in length and thickness and tend to fill available space more completely and uniformly. For example, the internal protoconid crest turns posteriad and encloses a less elongate, more oval fossettid than occurs in *Allomys nitens.* Most of these characters seem to represent a progressive condition over that of *Allomys nitens.* However, the smaller size and anteroposteriorly relatively shorter molars are similar to conditions in *Allomys simplicidens. Allomys reticulatus* may be derived from a descendant of *A. simplicidens* that paralleled *A. nitens* in crest development.

Allomys tessellatus occurs in the teilzone of *Entoptychus individens,* within but near the base of the Haystack Valley Member (figs. 45, 55). Only a single specimen has been recovered, so the vertical range is indeterminate. The crests of the lower molars in *A. tessellatus* tend to join one another to form almost twice as many complete fossettids as occur in *A. reticulatus* (pl. 11a). The sizes of the fossettids and of incompletely closed interspaces are rather constant, as in many specimens of *A. reticulatus.* This constancy and the resemblance in overall size suggest a relationship to and perhaps a stage of progression from *A. reticulatus:* for example, the internal protoconid crest curves posteriad and encloses an even more nearly circular fossettid than occurs in *A. reticulatus.* This and other specific changes such as the shortening of the anteroconid crest and the division of the central fossettid into two fossettids by the extension of an entoconid crest, are consistent with a derivation from *A. reticulatus* because these conditions are functionally or geometrically related to trends already present in that form. However one rather conspicuous character is not found in any of the earlier allomyine species. The lower cheek teeth are anteroposteriorly compressed with respect to width. Although the upper cheek teeth in both *A. reticulatus* and *A. simplicidens* are relatively shorter than in *A. nitens,* the lower cheek teeth in those taxa do not reflect that difference. The shortness of the lower cheek teeth in *A. tessellatus* seems to be an adaptation of different origin, because much of the shortness is caused by relatively short trigonids and a reduced hypoconulid lobe on M_3, characters not present in the other allomyines. The occurrence of *A. tessellatus* coincides with tectonic activity, the appearance of streams for the first time in the basin, and a pronounced faunal change involving appearances of camelids, mylagaulids and new geomyoids.

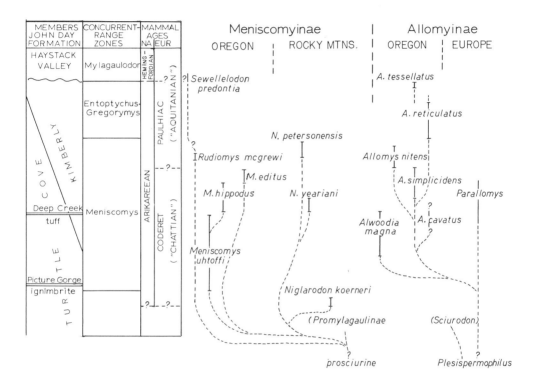

FIG. 55. Phylogenetic and geographic relationships of meniscomyine and allomyine taxa. The lower boundary of the *Meniscomys* Concurrent-range Zone is revised to a slightly earlier position with respect to the physical markers than that figured by Fisher and Rensberger (1972:fig. 6), based on the extended range of *Meniscomys* as described here. Boundary correlations of European mammal ages to North American mammal ages follow Van Couvering (1972). However, the primitive morphology of *Parallomys ernii* from the upper (Kuttigen) subdivision of the Coderet Age suggest that the top of the Coderet possibly correlates with the lower part of the *Meniscomys* Concurrent-range Zone and the Arikareean, or even earlier.

Intergeneric Relationships

The European genus *Parallomys* differs from *Allomys* in the generally greater simplicity of the occlusal surfaces and the less vertical, more strongly sloping labial faces of the paracone and metacone (pl. 6c,d). The interior of the lower molars is an almost smooth, open basin, broken only by very low and thin anteroconid and hypolophid crests. The central transverse valley of the upper cheek teeth is broader and more U-shaped than in *Allomys*. The simplest occlusal surfaces in *Allomys* are those of *A. cavatus,* which nevertheless contain more crests than do the surfaces in *Parallomys*. In *A. cavatus* (pl. 7a,b), a slightly thicker hypolophid crosses the center of the lower molars and small, accessory crests emanate from the protoloph and metaloph in the uppers. Although the stratigraphic position of *A. cavatus* is unknown, the most similar form, *A. simplicidens,* is the oldest in a stratigraphic series that acquires progressively more completely subdivided interior basins upward through the section (pls. 8–11a). Broad, relatively smooth occlusal basins therefore seem to characterize the primitive condition in the allomyines. Although *Parallomys* is in general the more primitive in appearance, the sloping condition of the faces of the paracone and metacone is less developed in North American allomyines and suggests that the ectoloph in the European group was already specialized in a somewhat different direction.

The lower dental structures of *Parallomys* are suggestive of derivation from a form like that of *Plesispermophilus*. The most striking aspects shared by these genera are the broad, open trigonid-talonid basin and the convex lingual margin in the lower molars, as well as the posteriorly produced, lingually expanded posterior lobe on M_3. These features are also shared with the primitive species of North American allomyines, and the posterior expansion of M_3 is retained through all but the most advanced species of *Allomys*. None of these conditions are shared with the known prosciurines of this country. However, the upper dentition of *Plesispermophilus* lacks the second, labial metaconule of the allomyines (Schmidt-Kittler and Vianey-Liaud, 1979:fig. 6). The resemblance of *Plesispermophilus* to *Sciurodon cadurcensis* in the incipient interior closure of the anterolabial molar fossettid and the absence of a labial metaconule suggest that *Sciurodon* may be derived from *Plesispermophilus* or a closely related taxon. The complete internal closure of the posterolingual fossettid (that is, complete posterior ectolophid) in *Plesispermophilus* differs from the internally open structure of *Parallomys,* but the primitive condition, present in all aplodontids of prosciurine grade, is that of internal closure. The available evidence therefore suggests that *Allomys* and *Parallomys* share a rather close common ancestry, that *Sciurodon* and

Plesispermophilus may be closely related, and that *Plesispermophilus* is closer to the ancestry of all these forms than any other known taxon, although not itself ancestral to *Parallomys* or *Allomys*.

Alwoodia seems somewhat more closely related to *Allomys* than to *Parallomys*, judging from the vertical labial faces of the paracone and metacone (pl. 12). However, it represents a distinct lineage. The advanced prominence of the major crests in *Alwoodia*, the tendency for almost parallel alignment of most of the crests in the lower molars, and the absence of small accessory crests (pls. 11c, 12a), indicate that this lineage had established a type of lophodonty distinct from that in *Allomys*. Even the earliest appearing species of *Allomys*, *A. simplicidens* (pl. 8), shows incipient accessory crests, and the later species (pls. 9, 10) never acquire such extended crests of the protoloph and metaloph as characterize *Alwoodia*.

In the prominence of the mesostylid and lingual convexity of the lower molars (pl. 11b), *Alwoodia* exhibits an aspect more like that of the European genus *Parallomys* than any of the species of *Allomys* except *A. cavatus*. *Alwoodia* therefore represents a group that had acquired heightened, strengthened and elongate internal crests early on but retained some very primitive structures shared with the European allomyine and lost in later North American forms.

The lower molars of *Meniscomys* differ from those of the allomyine genera in the presence of two prominent crests crossing the interior occlusal space, a mesostylid crest and a hypolophid (pl. 4a). *Meniscomys* also differs in the prominence of the mesostylid cusp and the deep notch that separates it from the metaconid (pl. 1a). The entoconid is positioned more posteriorly, in the posterolingual corner of the molars, the posterior margin is straighter, and the posterior lobe of M_3 is shorter (pl. 1c). The North American prosciurines, especially *Prosciurus,* are morphologically better candidates for the ancestry of *Meniscomys* than *Plesispermophilus,* because *Prosciurus* differs from *Plesispermophilus* in each of these characters except the development of the mesostylid crest. The crest connecting the mesostylid to the protoconid, the one structure unique to the meniscomyines, is absent in all described forms of prosciurine grade.

Niglarodon is closely related to *Meniscomys* but differs in the more widely open and larger posterolabial fossettid of the cheek teeth and the less bulbous, but longer, more lingually protrusive mesostylid in the molars (pl. 5b). The lower premolar in *Niglarodon* never acquired a mesostylid quite like that in *Meniscomys,* where in advanced forms it is usually a bulbous structure joined to the mesoconid by a crest (pl. 1d). However, a distinct premolar mesostylid is not primitive in either *Niglarodon* or *Meniscomys,* because the P_4 in primitive species of both groups is dominated by an elongate, bladelike metastylid crest in that region (pls. 2h, 5d).

The upper dentition of *Niglarodon* has been described only for two species. The P^3 in both these forms, from the Lemhi district of Idaho, is labially flatter than in any species of *Meniscomys*. The posterolabial fossette of P^4 is probably smaller in both forms than in *Meniscomys,* and the paracone seems to be anteroposteriorly relatively longer, resulting in an asymmetry. The occurrences of *Niglarodon* near the Lemhi River in eastern Idaho and those of *Meniscomys* in eastern Oregon represent the *Meniscomys* Concurrent-range Zone.

Rudiomys differs from both *Meniscomys* and *Niglarodon* in the arched condition of the mesostylid crest (pl. 5a), which wraps around a subcircular central fossettid on M_2, and from the allomyine genera in the presence of this crest. The mesostylid and metaconid are more widely separated (pl. 5c) than in either of the other meniscomyine genera; the metaconid is separated from metalophulid I by a notch (M_2 in pl. 5a), and the metastylid crest is more prominent than in the molars of other meniscomyines, though it is not as extended as in the allomyines (pl. 5a). The size of the metastylid crest is closer to that of a primitive promylagauline *Crucimys* (Rensberger, 1980) but extends posterolinguad rather than directly posteriad. The hypoconid is not posterolabially flattened as in promylagaulines. Although *Rudiomys* occurs stratigraphically above *Meniscomys,* it is more brachyodont than the promylagaulines or the other meniscomyines. The absence of vesicular septa and the retention of radial septa in the periotic of *Rudiomys* (pl. 5e) also place its divergence from primitive meniscomyines farther back than that of the known species of *Meniscomys.*

Sewelleladon shares several characteristics with *Rudiomys.* The metastylid crest is well developed, the mesostylid crest bends around a large central fossettid in the molars, and a notch separates the metaconid from metalophulid I (pl. 6a). Although *Sewelleladon* is much more advanced in hypsodonty (pl. 6b) and lophodonty, it occurs much higher in the stratigraphic section. It is possible that *Rudiomys* represents the group that gave rise to *Sewelleladon,* but the absence of a complete lower dentition of *Rudiomys* or an upper dentition of either form leave the relationship somewhat uncertain.

Conclusions

1. Two divergent major groups of nonprosciurine aplodontids occur in the John Day Formation. The Meniscomyinae are characterized by lower molars with a heavy transverse crest extending from the mesostylid to the protoconid, a prominent hypolophid and a rather well defined central fossettid between the two major crests. The Allomyinae primitively bear molars with essentially a single, unsubdivided basin, which in advanced forms becomes partitioned by numerous thin crests rather than a few heavy ones.

2. The morphology of the meniscomyine molars resembles that of the North American prosciurines in the strong hypolophid, distinct mesostylid, posterolingual prominence of the entoconid (giving the tooth an angular corner) and the posteriorly restricted hypoconid lake of M_3.

3. The primitive allomyine molar morphology more closely resembles that of the primitive European aplodontid, *Plesispermophilus*, than that of the North American prosciurines.

4. At least three successive species of *Meniscomys* occurred in the John Day Formation (youngest to oldest):

> *M. editus*, n. sp.
> *M. hippodus* Cope
> *M. uhtoffi*, n. sp.

5. The successive species of *Meniscomys* represent a morphologic progression involving structures related to hypsodonty and lophodonty, with the higher occurring forms being more hypsodont and more lophodont. Dentinal tracts become extended in the advanced species of *Meniscomys* but occur on the anterior and posterior sides of the cheek teeth rather than on the labial and lingual sides, as in *Entoptychus*, owing to the almost 90–degree difference in direction of occlusal motion in these genera.

6. *Meniscomys hippodus* is the most abundantly represented of the three species in all regions of its occurrence. It is also the most widespread species. Intraspecific morphologic differences between stratigraphic intervals at a single locality and between localities do not correlate with the overall trends toward increased hypsodonty and lophodonty that characterize the history of the genus.

7. *Meniscomys editus*, the terminal species, existed in smaller numbers than *M. hip-*

podus, unless factors affecting preservation declined temporarily over a wide region.

8. *Meniscomys* is not known in the Rocky Mountains or eastward. Species previously referred to *Meniscomys* from the Rocky Mountains and Great Plains represent distinct lineages and genera.

9. *Rudiomys mcgrewi* (n. gen., n. sp.) occurs high in the *Meniscomys* Concurrent-range Zone and represents a reappearance of the Meniscomyinae after the extinction of *Meniscomys.*

10. *Rudiomys mcgrewi* is more brachyodont than the most primitive species of *Meniscomys* and lacks the advanced stage of reticular partitioning of the bullar wall characteristic of *Meniscomys.*

11. The arched mesostylid crest and incipiently extended metastylid crest on the lower molars of *Rudiomys* suggest a possibly closer relationship to *Sewelleladon* than to *Meniscomys.*

12. *Sewelleladon* probably occurs either high in the *Entoptychus-Gregorymys* Concurrent-range Zone or in the *Mylagaulodon* Concurrent-range Zone.

13. Although *Sewelleladon* shares with the allomyines a continuous metastylid crest from metaconid to mesostylid, development of the other crests, hypsodonty, size of P_4 and the position and closure of the fossettids ally it more closely with the Meniscomyinae.

14. *Meniscomys petersonensis* Nichols, from near the boundary of the *Meniscomys* Concurrent-range Zone and the *Entoptychus-Gregorymys* Concurrent-range Zone in the Lemhi River drainage of Idaho, is more closely related to *Niglarodon koerneri* than to any of the species of *Meniscomys.* The distinct mesostylid on the P_4 of *Niglarodon petersonensis* is not a primitive feature in either *Meniscomys* or *Niglarodon* and was therefore independently acquired.

15. The isolated P_4's referred to *Meniscomys yeariani,* which is known by only the upper dentition, share mesostylid structure and relative anteroposterior elongation with *N. petersonensis.* Although these teeth represent a distinct species as Nichols (1976) concluded, they probably represent *Niglarodon,* not *Meniscomys.*

16. The skull, UM 5101, from the Lemhi region is characterized by a very large P^3 and the absence of an anterior cingulum on P^4, conditions not known in any other meniscomyine. These features are functionally related and are probably not merely anomalies of some other known form. With only one upper dentition of *Niglarodon* described, it is difficult to determine the relationship of UM 5101 to this widespread and probably diverse group.

17. The allomyines are represented in the John Day Formation by a temporal succession of at least five species, as follows (youngest to oldest):

Allomys tessellatus, n. sp.
Allomys reticulatus, n. sp.
Allomys nitens Marsh
Allomys simplicidens, n. sp.
Alwoodia magna, n. gen., n. sp.

In addition to these species, *Allomys cavatus,* known only from a type that lacks a locality specification, may occur somewhere stratigraphically beneath the occurrence of *A. simplicidens,* judging from the simplicity of its crest pattern and the vertical trend toward greater complexity.

18. *Alwoodia magna,* the earliest appearing species, is more advanced in crest thickness and height than any of the later allomyine species and could not have given rise to the latter. On the other hand, the lingual margin of the lower molars in *Alwoodia magna* are more convex than in any other species that locally succeeded it (but not more so than in *Allomys cavatus),* and small accessory crests are absent in the upper molars. These two features are probably primitive for allomyines because they characterize the very simple and very brachyodont molars of the European *Parallomys ernii* and are progressively modified in the North American sequence of *Allomys.*

19. In the sequence from *Allomys simplicidens* to *A. reticulatus,* internal crests of the molars become progressively higher. This change is most pronounced between *A. simplicidens* and *A. nitens.* From *A. nitens* to *A. reticulatus,* internal crests become more equidimensional: shorter and lower accessory crests become more consistently elongate and higher, and longer crests tend to become shorter.

20. In *A. tessellatus,* an increased number of the internal crests have become terminally fused to form a mosaic or netlike pattern of rather equidimensional fossettids.

21. Although differences in crest pattern between the successive species of *Allomys* generally form a progression, some characters suggest that *A. reticulatus* originated from some form related to *A. simplicidens* rather than from its immediate local predecessor, *A. nitens,* and that *A. tessellatus* also may have had a distant origin. *A. tessellatus* may have arrived in conjunction with a faunal change associated with the first appearance in the depositional basin of major stream systems and external drainage.

22. Some teeth recovered do not clearly fit into the named species, and the type of *A. cavatus* is not known from the local stratigraphic succession. These irregularities and conclusion 21 seem to imply that a greater diversity of species or subpopulations existed in or near the John Day basin than is represented by the described succession.

Appendices

Appendix 1
Abbreviations of Dental Characters

ACM^P	Depth of concavity on anterior surface of mesostyle on P^4, measured in direction perpendicular to posterior margin of tooth.
CF2	Degree of closure of central fossettid on M$_2$ (1 = not indicated; 2 = incomplete; 3 = completed by process connecting mesostylid crest and hypolophid).
CH^1	Chevron height on M^1, measured from upper margin of lingual enamel at base of protocone to apex of posterior dentinal tract.
CH^2	As CH^1, but for M^2.
CLP^P	Depth of curvature of lingual surface of paracone on P^4, measured in occlusal plane from chord extended between labialmost arms of anterolabial fossette.
DMM^3	Presence of double mesostyle on M^3; presence indicated by two narrow ribs on labial surface of ectoloph (0 = absent; 1 = present).
E^P	Height of lingual enamel on P^4, measured vertically from level of enamel base on anterior face of anterocone to margin of enamel at base of protocone.
HAI1	Height of base of anterolingual inflection above base of enamel at anterolingual corner of crown on M$_1$.
HAI2	As HAI1, but for M$_2$.
HALI^P	Height of enamel above base of anterolabial inflection (situated lingual to anterocone) on P^4.
HC	Height of crests above bottoms of valleys on cheek teeth (0 = low; 1 = high).
HGP^3	Prominence of groove(s) on posterior surface of P^3 (0 = absent; 1 = faint; 2 = prominent).
HMCP	Height of metaconid above enamel base on P$_4$.
HMSAP	Height of mesostylid above base of anterolingual inflection on P$_4$.
HMSP	Height of mesostylid above base of posterolingual inflection on P$_4$.
HPIP	Height of posterolingual inflection base above base of lingual enamel on P$_4$.
HPI1	As HPIP, but for M$_1$.
HPI2	As HPIP, but for M$_2$.
HP^3	Height of apex of P^3 above enamel base.
HSHP	Height of stylid on anterior surface of hypoconid on P$_4$.
IMC2	Degree of development of an internal accessory crest on the metaconid of M$_2$ (or M$_3$), that is, in addition to the metalophid.
IPC2	Direction of curvature of interior end of internal protoconid crest on M$_2$ (posterolabiad or posterolinguad).
LAC2	Anteroposterior length of anteroconid crest on M$_2$.
LI	Anteroposterior length of cross-sectional area of lower incisor, measured perpendicular to trend of enamel face.
LIAP	Vertical angle (in degrees) between walls of labial inflection on P$_4$, in labial view.

LLM^1	Anteroposterior length of lingual moiety of M^1, measured near base of crown above flared area.
LL^P	Anteroposterior length of lingual moiety of P^4, measured between bases of anterolingual and posterolingual inflections.
LM1	Anteroposterior length of M$_1$, measured at level of base of posterolingual inflection (meniscomyines) or in occlusal view (allomyines).
LM2	As LM1, but for M$_2$.
LM^1	Anteroposterior length of M^1, measured parallel to labial surface of anterior ectoloph (meniscomyines) or perpendicular to anterior cingulum (allomyines).
LM^2	As LM1, but for M^2.
LM^3	As LM1, but for M^3.
LPE^P	Anteroposterior length of posterior segment of ectoloph on P^4, measured between posterolabial corner of tooth and center of mesostyle at about level of base of posterolingual inflection.
LP	Anteroposterior length of P$_4$, measured at level of posterolingual inflection (meniscomyines) or in occlusal view (allomyines).
L^P	Anteroposterior length of P^4, measured parallel to labial surface of ectoloph.
LP^3	Anteroposterior length of P^3.
LTR1	Anteroposterior length of trigonid on M$_1$, measured from base of labial inflection forward.
LTR2	As LTR1, but for M$_2$.
MLC^1	Anteroposterior separation of bend of metaloph on M^1 from midline of central valley.
MP	Presence or absence of mesostylid process on mesoconid of P$_4$ (0 = absent; 1 = present).
MSS	Presence or absence of mesostylid on P$_4$ (0 = absent; 1 = present).
NF2	Number of fossettids on M$_2$.
NPC^1	Number of accessory crests on posterior surface of protoloph, protoconule and paracone on M^1.
NMC^1	Number of accessory crests on anterior surface of metaloph, metaconule and metacone of M^1.
PAPR^1	Size of anterior process on paracone of M^1 (0 = absent; 1 = small; 2 = large).
SHLP	Presence or absence of anterior spur on hypoconulid of P$_4$ (0 = absent; 1 = present).
TC^1	Total number of accessory crests (in addition to crests forming ectoloph, protoloph and metaloph) on M^1, emanating from paracone, metacone, protoconule, metaconule, labial metaconule, protocone and cingula.
TC^2	As TC^1, but for M^2.
TCDP^4	As TC^1, with addition of crests from anterocone, but for DP^4.
TEI	Maximum thickness of enamel on lower incisor.
WAP	Width of anterior moiety of P$_4$, measured at level of base of posterolingual inflection.
WAIP	Transverse width of anterolingual inflection on P$_4$, measured from lingual margin of mesostylid.
WCF^3	Transverse width of central fossette on M^3.
WI	Transverse width of lower incisor, measured parallel to trend of enamel face.
W^I	As WI, but for upper incisor.
WLL^P	Transverse width of lingual loph of moderately worn P^4, viewed parallel to lingual surface and measured from anteroposterior line through lingual end of protoloph to anteroposterior line tangent to lingual margin of crown.

WM^1 Transverse width of M^1, measured from lingual margin of crown in direction perpendicular to ectoloph (meniscomyines) or parallel to anterior cingulum (allomyines).

WM^2 As WM^1 but for M^2.

WPL^P Transverse width of parastyle and anterocone on P^4, measured from labial margin of rib to anterolabial inflection.

WPP   Transverse width of posterior moiety of P$_4$, measured at level of base of posterolingual inflection.

W^P Transverse width of P^4.

WP^3 Transverse width of P^3.

WTR1   Transverse width of trigonid on M$_1$, measured at level of maximum flare near base of protoconid.

WT2   Transverse width of talonid on M$_2$, measured at level of base of posterolingual inflection.

Appendix 2
Localities

Haystack 1   (V–6429). Center, sec. 21, R. 25 E., T. 8 S., Kimberly quadrangle, Oregon, 1953. Fossils from green, zeolitic siltstone of upper part of Turtle Cove Member, John Day Formation.

Haystack 6   (V–6590, UWA 4799). NE ¼ NW ¼ sec. 4, R. 25 E., T. 9 S., Kimberly quadrangle, Oregon, 1953. Green, zeolitic siltstone of Turtle Cove Member, John Day Formation.

Haystack 6A   (V–6505). Center, SW ¼ sec. 34, R. 25 E., T. 8 S., Kimberly quadrangle, Oregon, 1953. Green, zeolitic siltstone of Turtle Cove Member, John Day Formation.

Haystack 8   (V–6322). Center, N ½ NE ¼ sec. 34, R. 25 E., T. 8 S., Kimberly quadrangle, Oregon, 1953. Green, zeolitic siltstone of Turtle Cove Member, John Day Formation.

Haystack 19   (V–6587). SE side of creek, SE ¼ NE ¼ sec. 33, R. 25 E., T. 8 S., Kimberly quadrangle, Oregon, 1953. Greenish sandstone, siltstone of Haystack Valley Member, John Day Formation.

Haystack 32   (V–6581, UWA 5846). S. of hill 2976, secs. 15, 16, 21, 22, R. 25 E., T. 8 S., Kimberly quadrangle, Oregon, 1953. Green, zeolitic siltstone of Turtle Cove Member, John Day Formation.

Haystack 33   (V–66104). SW ¼ NW ¼ sec. 22, R. 25 E., T. 8 S., Kimberly quadrangle, Oregon, 1953. Green, zeolitic siltstone of Turtle Cove Member, John Day Formation.

Hinton   (UWB 1039). NW ¼ SE ¼ sec. 28, R. 28 E., T. 9 S. Monument quadrangle, Oregon, 1953. *Meniscomys* from gray siltstone, 3 m beneath lower resistant tuff, Turtle Cove Member, John Day Formation.

Picture Gorge 6   (V–6662, UWA 9581). SW ¼ NE ¼ sec. 18, R. 26 E., T. 10 S., Picture Gorge quadrangle, Oregon, 1953. Green, zeolitic siltstone of Turtle Cove Member, John Day Formation.

Picture Gorge 7   (V–6506, UWA 5183). Center, NE ¼ sec. 19, R. 26 E., T. 10 S., Picture Gorge quadrangle, Oregon, 1953. Green, zeolitic siltstone of Turtle Cove Member, John Day Formation.

Picture Gorge 12   (V–6685, UWA 9591). NE ¼ NW ¼ sec. 36, R. 25 E., T. 10 S., Picture Gorge quadrangle, Oregon, 1953. Fossils from green, zeolitic siltstone of Turtle Cove Member, John Day Formation, above Picture Gorge ignimbrite.

Picture Gorge 17   (V–66111, UWA 5171). NW ¼ NW ¼ sec. 36, R. 25 E., T. 10 S., Picture Gorge quadrangle, Oregon, 1953. Fossils from green, zeolitic siltstone of Turtle Cove Member, John Day Formation, above Deep Creek tuff.

Picture Gorge 19	(V–66113, UWA 9592). NW ¼ NW ¼ sec. 34, R. 25 E., T. 10 S., Picture Gorge quadrangle, Oregon, 1953. Green, zeolitic siltstone of turtle Cove Member, John Day Formation.
Picture Gorge 20	(UWA 4556). Center SE ¼ sec. 31, R. 26 E., T. 10 S., Picture Gorge quadrangle, Oregon, 1953. Level 2 close beneath Picture Gorge ignimbrite. Green, zeolitic siltstone.
Picture Gorge 22	(V–66116, UWA 5172). From center to eastern boundary sec. 1, R. 25 E., T. 11 S., Picture Gorge quadrangle, Oregon, 1953. Green, zeolitic siltstone, Turtle Cove Member, John Day Formation, beneath Deep Creek tuff.
Picture Gorge 29	(V–6649, UWA 9596). SE ¼ SE ¼ sec. 15, R. 26 E., T. 11 S., Picture Gorge quadrangle, Oregon, 1953. *Allomys* and *Meniscomys* from buff colored siltstone of Turtle Cove Member, John Day Formation, above green siltstone.
Picture Gorge 33	(V–66123, UWA 5833). Center, E. ½ SE ¼ sec. 7, R. 26 E., T. 12 S., Picture Gorge quadrangle, Oregon, 1953. Pale green to buff colored zeolitic siltstone of Turtle Cove Member, John Day Formation.
Picture Gorge 34	(V–66124, UWA 5834). Center, E. edge NW ¼ sec. 7, R. 26 E., T. 12 S., Picture Gorge quadrangle, Oregon, 1953. Pale green to buff colored siltstone of Turtle Cove Member, John Day Formation.
Picture Gorge 43	(UWA 9966). NW ¼ NW ¼ NW ¼ sec. 8, R. 26 E., T. 12 S., Picture Gorge quadrangle, Oregon, 1953. *Allomys* from base of buff colored siltstone of Turtle Cove Member, John Day Formation, above Picture Gorge ignimbrite.
Rudio Creek 2	(V–7077). Center, N. ¼ NE ¼ sec. 27, R. 26 E., T. 9 S., Kimberly quadrangle, Oregon, 1953. Pale gray siltstone of Kimberly Member, John Day Formation.
Rudio Creek 3	(V–66106). NE ¼ SE ¼ sec. 26, R. 26 E., T. 9 S., Kimberly quadrangle, Oregon, 1953. Light gray siltstone of Kimberly Member, John Day Formation.
Rudio Creek 4	(V–6600, UWA 5929). Center, W. ½ SE ¼ sec. 22, R. 26 E., T. 9 S., Kimberly quadrangle, Oregon, 1953. Light gray siltstone of Kimberly Member, John Day Formation.
Schrock's 1	(V–6351). W. ½ sec. 19, R. 2l E., T. 18 S., Crescent (1:250,000) sheet, Oregon, 1958. Gray or greenish siltstone of Kimberly Member, John Day Formation.
Stubblefield 1A	(V–6658). SW ¼ NW ¼ sec. 28, R. 28 E., T. 9 S., Monument quadrangle, Oregon, 1953. Gray unzeolitized siltstone of Kimberly Member, John Day Formation.
Weaver's	(V–6645). Secs. 16, 22, R. 21 E., T. 19 S., Crescent (1:250,000) sheet, Oregon, 1958. *Meniscomys* from green, zeolitic siltstone of Turtle Cove Member, John Day Formation, above Deep Creek tuff.

Literature Cited

Black, C. C.
1961. Rodents and lagomorphs from the Miocene Fort Logan and Deep River formations of Montana. Postilla 48, 20 p.
1969. The fossil rodent genera *Horatiomys* and *Palustrimus* — juvenile geomyoid rodents. J. Mamm. 50:815–817.

Butler, P. M.
1952. Molarization of the premolars in the Perissodactyla. Proc. Zool. Soc. London 121:777–817.

Cita, M. B.
1968. Report of the working group, micropaleontology. C. M. N. S. Proc. Sess. IV, Bologna, 1967; Gior. Geol. 35 (II):21–22.

Cope, E. D.
1879. On some characters of the Miocene fauna of Oregon. Proc. Amer. Phil. Soc. 18:63–78.
1881. Review of the Rodentia of the Miocene Period of North America. Bull. U.S. Geol. Geog. Surv. Terr. 6:361–386.
1884. The Vertebrata of Tertiary formations of the west. Rept. U.S. Geol. Surv. Terr. 3, 1009 p.

Dehm, R.
1950. Die Nagetiere aus dem Mittel-Miocan (Burdigalium) von Wintershof-West bei Eichstatt in Bayern. N. Jahrb. f. Miner. Geol. Palaeontol. 91:321–428.

Fisher, R. A.
1970. Statistical methods for research workers. Hafner Press, New York. 362 p.

Fisher, R. V.
1962. Clinoptilolite tuff from the John Day Formation, eastern Oregon. The Ore Bin 24:197–203.
1963. Zeolite-rich beds of the John Day Formation, Grant and Wheeler counties, Oregon. The Ore Bin 25:185–197.
1964. Iron oxide rings probably result of fumarole activity in Miocene ignimbrite. Geol. Soc. Amer. Spec. Paper 76:58.
1966. Geology of a Miocene ignimbrite layer, John Day Formation, eastern Oregon. Univ. Calif. Publ. Geol. Sci. 67, 73 p.
1967. Early Tertiary deformation in north-central Oregon. Amer. Assoc. Petrol. Geol. Bull. 51:111–123.

Fisher, R. V. and J. M. Rensberger
1972. Physical stratigraphy of the John Day Formation, central Oregon. Univ. Calif. Publ. Geol. Sci. 101, 45 p.

Furlong, E.
1910. An aplodont rodent from the Tertiary of Nevada. Univ. Calif. Bull. Dept. Geol. 3:397–403.

George, T. N., and members of the Stratigraphy Committee
1969. Recommendations on stratigraphical usage. Proc. Geol. Soc. London 1656:139–166.

Hay, R. L.
1963. Stratigraphy and zeolitic diagenesis of the John Day Formation of Oregon. Univ. Calif. Publ. Geol. Sci. 42:199–262.

Hugueney, M.
1969. Les rongeurs (Mammalia) de l'Oligocene superieur de Coderet-Bransat (Allier). Publ. Ph.D. thesis, Univ. Lyon. 238 p.

Klingener, D.
  1968. Rodents of the Mio-Pliocene Norden Bridge Local Fauna, Nebraska. Amer. Midland Nat. 80:65–74.
Macdonald, L. J.
  1972. Monroe Creek (early Miocene) microfossils from the Wounded Knee area, South Dakota. South Dakota Geol. Surv. Rept. Invest. 105, 43 p.
Macdonald, J. R.
  1963. The Miocene faunas from the Wounded Knee area of western South Dakota. Bull. Amer. Mus. Nat. Hist. 125:141–238.
  1970. Review of the Miocene Wounded Knee faunas of southwestern South Dakota. Bull. Los Angeles Co. Mus. Nat. Hist. Sci. 8, 82 p., 2 maps.
Marsh, O. C.
  1877. Notice of some new vertebrate fossils. Amer. J. Sci., ser. 3, 14:249–256.
McGrew, P. O.
  1941. The Aplodontoidea. Field Mus. Nat. Hist., Geol. Ser. 9:1–30.
Merriam, J. C.
  1900. Classification of the John Day beds. Science, n.s., 11:219–220.
  1901a. A contribution to the geology of the John Day basin. Univ. Calif. Bull. Dept. Geol. 2:269–314.
  1901b. Geological section through the John Day basin (abs.) Bull. Geol. Soc. Amer. 12:496–497.
Merriam, J. C. and W. J. Sinclair
  1907. Tertiary faunas of the John Day region. Univ. Calif. Bull. Dept. Geol. 5:171–205.
Miller, G. S. Jr. and J. W. Gidley
  1918. Synopsis of the supergeneric groups of rodents. J. Wash. Acad. Sci. 8:431–448.
Nichols, R.
  1976. Early Miocene mammals from the Lemhi Valley of Idaho. Tebiwa 18:9–47.
Rensberger, J. M.
  1971. Entoptychine pocket gophers (Mammalia, Geomyoidea) of the early Miocene John Day Formation, Oregon. Univ. Calif. Publ. Geol. Sci. 90, 209 p.
  1973a. An occlusion model for mastication and dental wear in herbivorous mammals. J. Paleontol. 47:515–528.
  1973b. Pleurolicine rodents (Geomyoidea) of the John Day Formation, Oregon, and their relationships to taxa from the early and middle Miocene, South Dakota. Univ. Calif. Publ. Geol. Sci. 102, 131 p.
  1975a. Function in the cheek tooth evolution of some hypsodont geomyoid rodents. J. Paleontol. 49:10–22.
  1975b. *Haplomys* and its bearing on the origin of the aplodontoid rodents. J. Mamm. 56:1–14.
  1979. *Promylagaulus*, progressive aplodontoid rodents of the early Miocene. Los Angeles Co. Mus. Nat. Hist., Contr. Sci. 312, 18 p.
  1980. A primitive promylagauline rodent from the Sharps Formation, South Dakota. J. Paleontol. 54:1267–1277.
  In   Evolution in a late Oligocene-early Miocene succession of meniscomyine rodents in the
  press. Deep River Formation, Montana. J. Vertebrate Paleontol.
Savage, D. E., D. E. Russell, and P. Louis
  1965. European Eocene Equidae (Perissodactyla). Univ. Calif. Publ. Geol. Sci. 56, 98 p.
Schmidt-Kittler, N. and M. Vianey-Liaud
  1979. Evolution des Aplodontidae Oligocenes Europeens. Palaeovertebrata 9:31–82.

Shotwell, J. A.
1958. Evolution and biogeography of the aplodontid and mylagaulid rodents. Evol. 12:451–484.

Stehlin, H.G. and S. Schaub
1951. Die Trigonodontie der simplicidentaten Nager. Schweiz. Palaontol. Abh. 67, 385 p.

Thaler, L.
1966. Les Rongeurs fossiles du Bas-Languedoc dans leurs rapports avec l'histoire des faunes et la stratigraphie du Tertiare d'Europe. Mem. Mus. Nat. d'Hist. Natur., n.s., 17, 296 p.

Van Couvering, J. A.
1972. Radiometric calibration of the European Neogene. *In* Bishop, W. W. and J. A. Miller, eds., Calibration of hominoid evolution, Scottish Acad. Press, p. 247–271.

Viret, J. and M. Casoli
1961. Sur un Rongeur de la famille americaine des Aplodontides decouvert dans le Stampien superieur de la Limagne bourbonnaise. Eclogae geol. Helv. 54:541–545.

Wood, A. E.
1935. Two new genera of cricetid rodents from the Miocene of western United States. Amer. Mus. Novitates 789, 8 p.
1936. Geomyid rodents from the middle Tertiary. Amer. Mus. Novitates 866, 31 p.

Plates

PLATE 1

a, Meniscomys uhtoffi, left mandible, P_4-M_3, lingual view, UCMP 76514, Picture Gorge 12 (V–6685).

b, Meniscomys hippodus, right mandible, P_4-M_3, lingual view, UWBM 29157, Picture Gorge 7 (UWA 5183), level 1.

c, Meniscomys uhtoffi, left P_4-M_3, occlusal view, UCMP 76514, Picture Gorge 12 (V–6685), stereo pair.

d, Meniscomys hippodus, right P_4-M_3, occlusal view, UWBM 29157, Picture Gorge 7 (UWA 5183), level 1, stereo pair. White square = 1 mm × 1 mm.

Plate 1 135

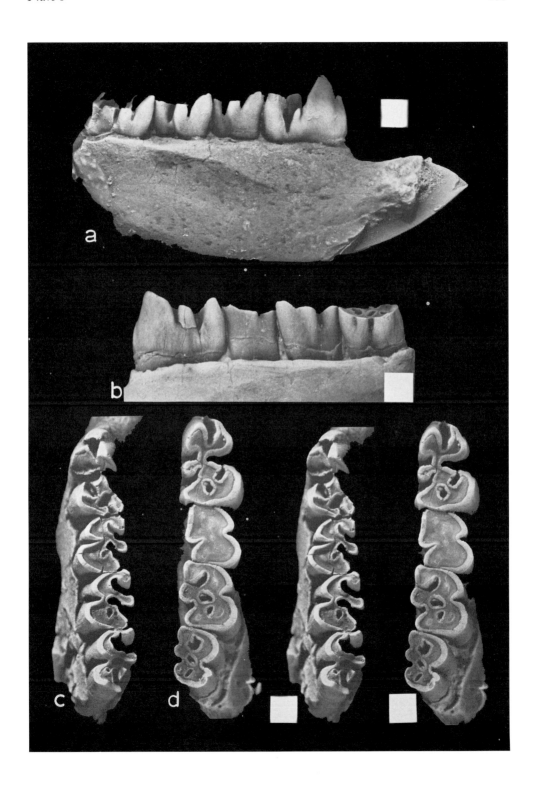

PLATE 2
Meniscomys hippodus:

   a, left DP₄-M₁, occlusal view, UWBM 39512, Picture Gorge 17 (UWA 5171), stereo pair.

   b, left M₂, lingual view, anterior right, UWBM 43374, Picture Gorge 17 (UWA 5171), stereo pair.

   c, same, occlusal view, anterior left, stereo pair.

   d, left DP₄-M₁, lingual view, anterior right, UWBM 39512, Picture Gorge 17 (UWA 5171).

   e, right P₄, labial view, anterior right, UWBM 29357, Picture Gorge 7 (UWA 5183), level 2.

   f, same, occlusal view, anterior left, stereo pair.

   g, same, lingual view.

Meniscomys uhtoffi:

   h, right P₄, lingual view, anterior left, UWBM 43024, Picture Gorge 12 (UWA 9591), stereo pair.

   i, same, occlusal view, stereo pair. White square = 1 mm × 1 mm.

Plate 2 137

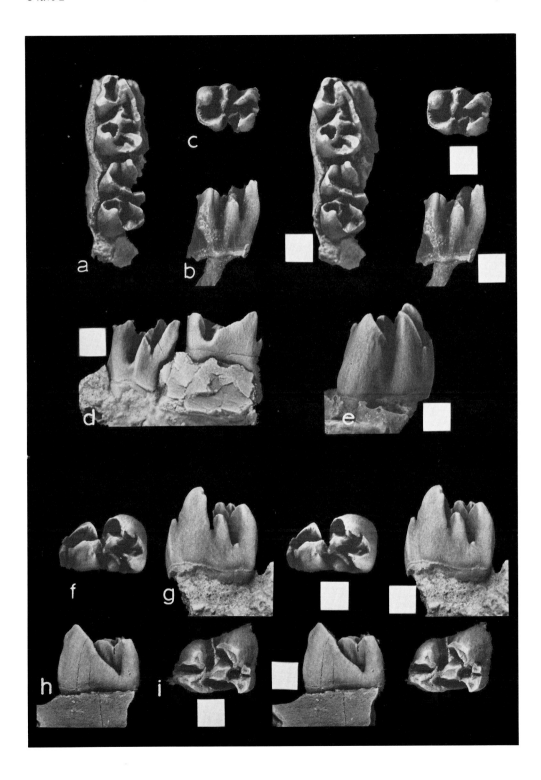

PLATE 3

Meniscomys hippodus:

a, left P⁴-M¹, occlusal view, UWBM 29233, Picture Gorge 7 (UWA 5183), level 2, stereo pair.

b, left P³-M³, occlusal view, UCMP 76803, Picture Gorge 7 (V–6506), level 2, stereo pair.

c, right tympanic, ventral view, anterior top, UCMP 105021, Haystack 33 (V–66104), stereo pair.

d, right DP⁴-M¹, occlusal view, anterior right, UWBM 43379, Picture Gorge 17 (UWA 5171), stereo pair.

Meniscomys editus:

e, left P₄-M₃, lingual view, anterior right, UWBM 39557, Picture Gorge 7 (UWA 5183), level 2 (+). White square = 1 mm × 1 mm.

Plate 3 139

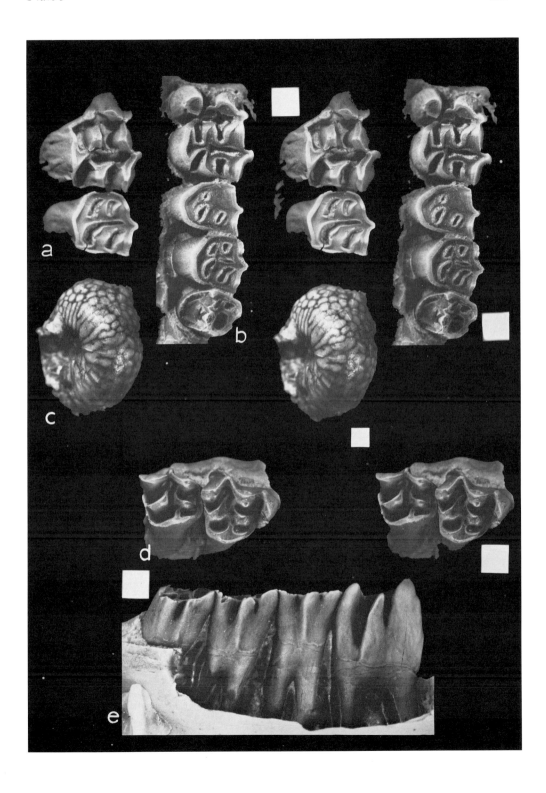

PLATE 4
Meniscomys editus:

a, left P$_4$-M$_3$, occlusal view, UWBM 39557, Picture Gorge 7 (UWA 5183), level 2 (+), stereo pair.

b, right P^3-M^3, occlusal view, UWBM 54900, Picture Gorge 7, level 3, stereo pair.

c, right P$_4$-M$_3$, occlusal view, UWBM 54900, stereo pair. White square = 1 mm × 1 mm.

Plate 4 141

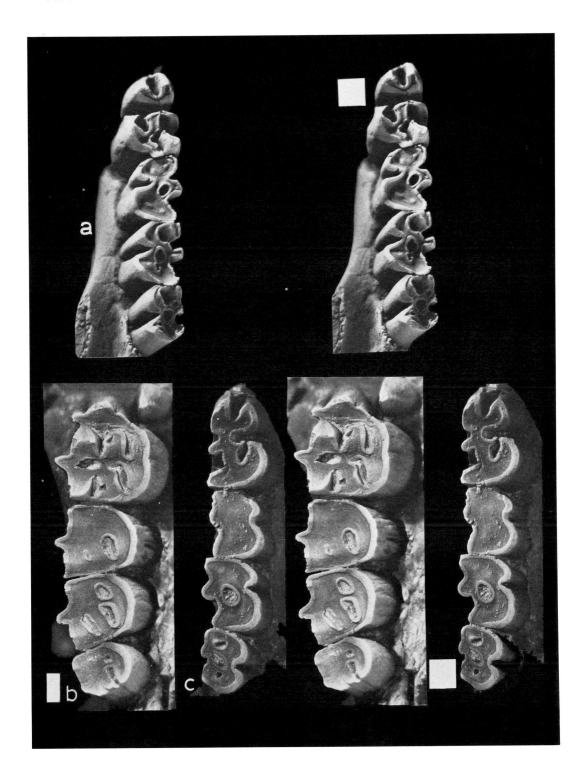

PLATE 5

a, Rudiomys mcgrewi, left M$_{1-2}$, occlusal view, UCMP 105122, Rudio Creek 3 (V–66106), stereo pair.

b, Niglarodon koerneri, right P$_4$-M$_3$, occlusal view, YPM 14024, Meagher County, Montana, stereo pair.

c, Rudiomys mcgrewi, left M$_{1-2}$, lingual view, anterior right, UCMP 105122, Rudio Creek 3 (V–66106).

d, Niglarodon koerneri, right P$_4$-M$_3$, lingual view, anterior left, YPM 14024, Meagher County, Montana.

e, Rudiomys mcgrewi, left tympanic, ventrolateral view, anterior top, UCMP 105122, Rudio Creek 3 (V–66106), stereo pair. White square = 1 mm × 1 mm.

Plate 5 143

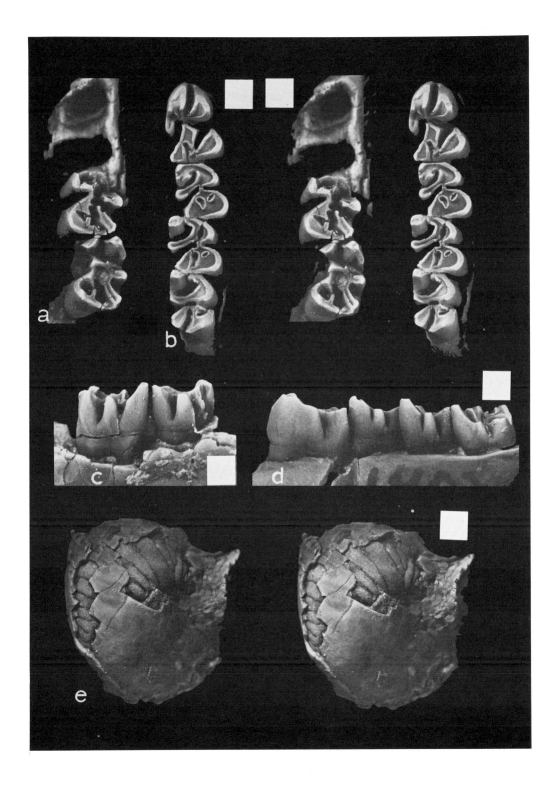

PLATE 6

a, Sewelleladon predontia, left P₄-M₃, occlusal view, UOMNH F-4734, locality UO 2275, stereo pair.

b, same, lingual view.

c, Parallomys ernii, left M¹ (cast), occlusal view, anterior left, UL 96305, Coderet quarry, France, stereo pair.

d, Parallomys ernii, left M₁ (cast, bubble on apex of metaconid), occlusal view, UL 96303, Coderet quarry, France, stereo pair. White square = 1 mm × 1 mm.

Plate 6

145

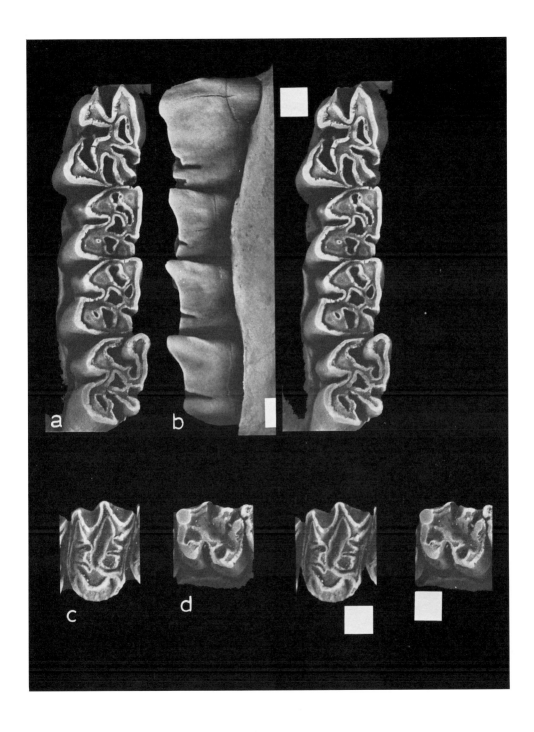

PLATE 7

Allomys cavatus, AMNH 6988:

a, left P^4-M^3 (cast), occlusal view, stereo pair.

b, left M$_{1-3}$ (cast), occlusal view, stereo pair.

c, tympanics and basicranium (cast), showing natural endocast and septa, in part; stereo pair.
White square = 1 mm × 1 mm.

Plate 7

147

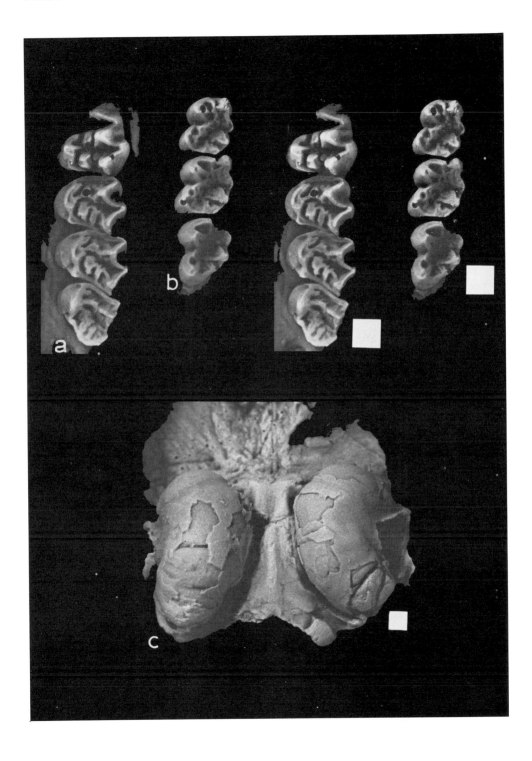

PLATE 8
Allomys simplicidens:

a, left P³-M², occlusal view, UWBM 29156, Picture Gorge 7 (UWA 5183), level 1, stereo pair.

b, left M₂, occlusal view, anterior top, UWBM 29297, Picture Gorge 7 (UWA 5183), level 2, stereo pair.

c, right P₄, occlusal view, anterior top, UWBM 29228, Picture Gorge 7 (UWA 5183), level 1, stereo pair.

d, right M₁₋₂, occlusal view, anterior top, UWBM 54800, Picture Gorge 7 (UWA 5183), level 3, stereo pair.

e, same, lingual view, stereo pair.

f, left DP₄, occlusal view, anterior left, UCMP 75922, Schrock's 1 (V-6351), level O, stereo pair. White square = 1 mm × 1 mm.

Plate 8 149

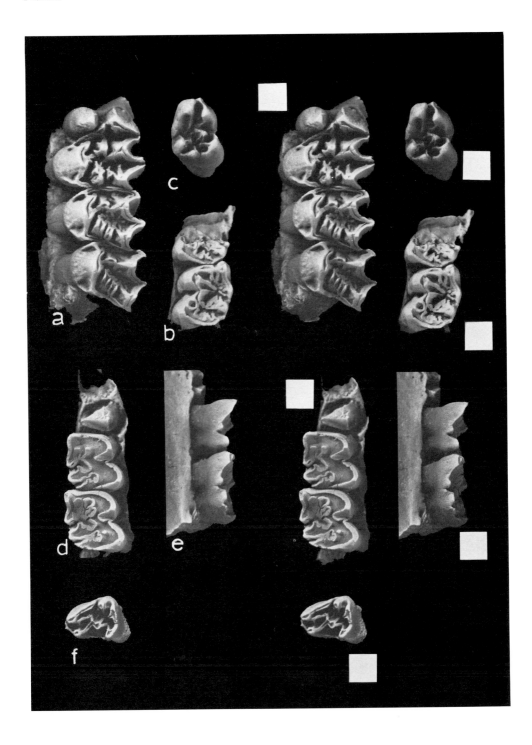

PLATE 9
Allomys nitens:

a, left P³-M², occlusal view, YPM 13604, stereo pair.

b, left DP⁴-M², occlusal view, UCMP 107717, Picture Gorge 33 (V–66123), stereo pair.

c, left P₄-M₃, occlusal view, UWBM 58058, Picture Gorge 33 (UWA 5833), stereo pair.

d, same, lingual view, stereo pair. White square = 1 mm × 1 mm.

Plate 9 151

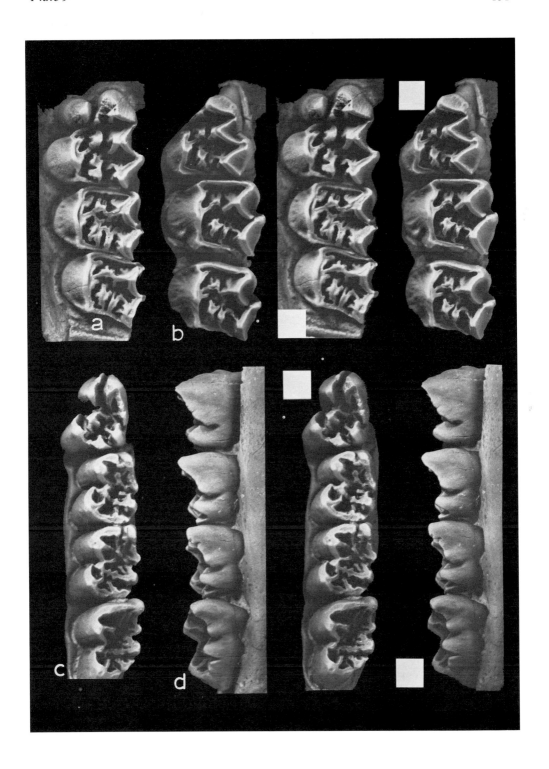

PLATE 10

a, Allomys nitens, right P$_4$-M$_1$, occlusal view, UWBM 51967, Picture Gorge 34 (UWA 5834), stereo pair.

b, Allomys nitens, left M$_2$-$_3$, occlusal view, UWBM 31473, Picture Gorge 34 (UWA 5834), stereo pair.

c, Allomys nitens, left M$_1$-$_3$, occlusal view, UCMP 97068, Haystack 6 (V–6590), stereo pair.

d, Allomys reticulatus, left M$_1$-$_3$, occlusal view, UCMP 75894, Schrock's 1 (V–6351), level 8, stereo pair.

e, Allomys reticulatus, left P$_4$-M$_2$, occlusal view, UCMP 75942, Schrock's 1 (V–6351), level 8, stereo pair.

f, Allomys reticulatus, right P^3-M^2, occlusal view, UCMP 105039, Haystack 6A (V–6505), stereo pair. White square = 1 mm × 1 mm.

Plate 10

153

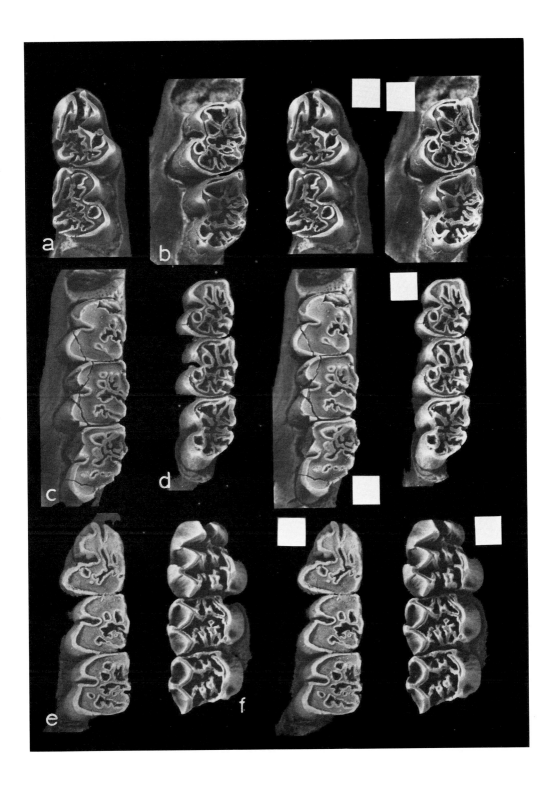

PLATE 11

a, Allomys tessellatus, left M$_{1-3}$, occlusal view, UCMP 105038, Haystack 19 (V–6587), stereo pair.

b, Alwoodia magna, left M$_{1-2}$, occlusal view, UWBM 43035, Picture Gorge 12 (UWA 9591), stereo pair.

c, Alwoodia magna, left P$_4$-M$_3$, occlusal view, UWBM 47336, Picture Gorge 29 (UWA 9596), level 2, stereo pair.

d, same, lingual view, stereo pair. White square = 1 mm × 1 mm.

Plate 11   155

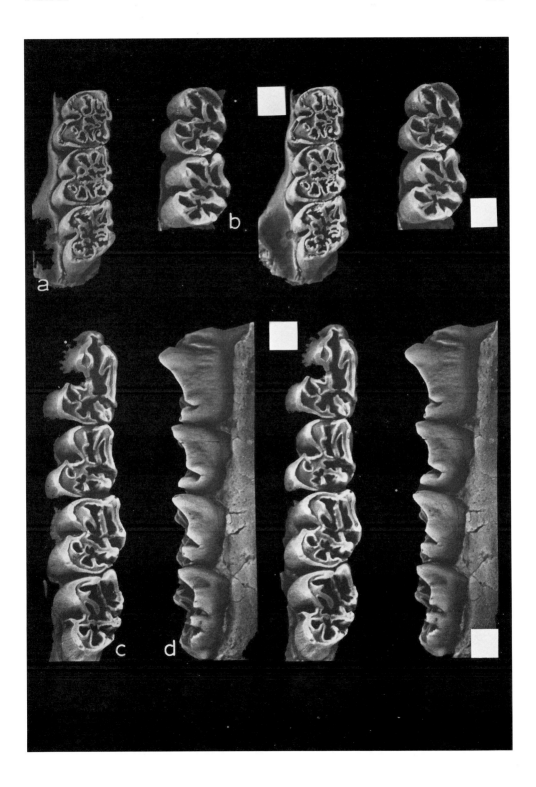

PLATE 12

Alwoodia magna:

a, right P³-M³, occlusal view, UCMP 76941, Picture Gorge 22 (V–66116), stereo pair.

b, left DP³-⁴, M¹-³, occlusal view, UCMP 76938, Picture Gorge 22 (V–66116), stereo pair.

c, skull, ventral view , UCMP 76938, Picture Gorge 22 (V–66116). White square = 1 mm × 1 mm.

Plate 12   157

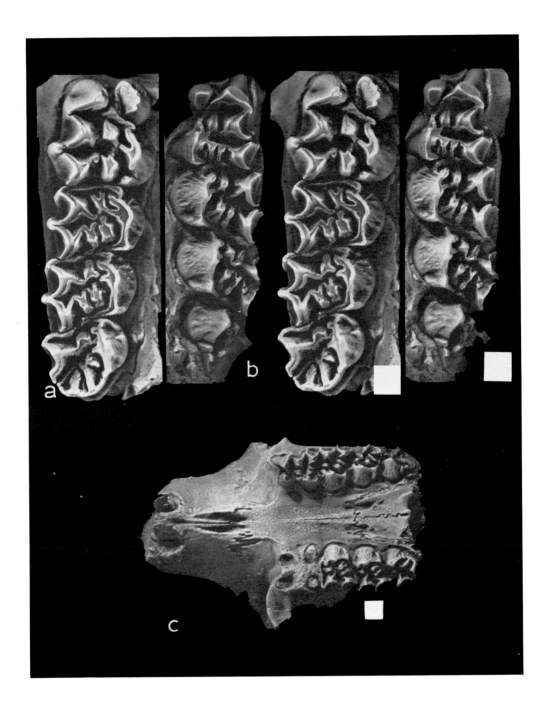